# Generative Adversarial Networks and Deep Learning

This book explores how to use generative adversarial networks in a variety of applications and emphasises their substantial advancements over traditional generative models. This book's major goal is to concentrate on cutting-edge research in deep learning and generative adversarial networks, which includes creating new tools and methods for processing text, images, and audio.

A generative adversarial network (GAN) is a class of machine learning framework and is the next emerging network in deep learning applications. Generative adversarial networks (GANs) have the feasibility to build improved models, as they can generate the sample data as per the application requirements. There are various applications of GAN in science and technology, including computer vision, security, multimedia and advertisements, image generation, image translation, text-to-images synthesis, video synthesis, generating high-resolution images, drug discovery, etc.

Features:

- Presents a comprehensive guide on how to use GAN for images and videos
- Includes case studies of underwater image enhancement using generative adversarial network, intrusion detection using GAN
- Highlights the inclusion of gaming effects using deep learning methods
- Examines the significant technological advancements in GAN and its real-world application
- Discusses GAN challenges and optimal solutions

The book addresses scientific aspects for a wider audience such as junior and senior engineering, undergraduate and postgraduate students, researchers, and anyone interested in the trends development and opportunities in GAN and deep learning.

The material in the book can serve as a reference in libraries, accreditation agencies, government agencies, and especially the academic institutions of higher education intending to launch or reform their engineering curriculum.

# Generative Adversarial Networks and Deep Learning

## Theory and Applications

Edited by

Roshani Raut
Pranav D Pathak
Sachin R Sakhare
Sonali Patil

CRC Press is an imprint of the
Taylor & Francis Group, an **informa** business
A CHAPMAN & HALL BOOK

First edition published 2023
by CRC Press
6000 Broken Sound Parkway NW, Suite 300, Boca Raton, FL 33487-2742

and by CRC Press
4 Park Square, Milton Park, Abingdon, Oxon, OX14 4RN

*CRC Press is an imprint of Taylor & Francis Group, LLC*



*Library of Congress Cataloging-in-Publication Data*
Names: Raut, Roshani, 1981– editor.
Title: Generative adversarial networks and deep learning : theory and applications /
edited by Roshani Raut, Pranav D Pathak, Sachin R Sakhare, Sonali Patil.
Description: First edition. | Boca Raton : Chapman & Hall/CRC Press, 2023. |
Includes bibliographical references and index. |
Summary: "This book explores how to use generative adversarial networks (GANs) in a variety of applications and emphasises their substantial advancements over traditional generative models. This book's major goal is to concentrate on cutting-edge research in deep learning networks and GANs, which includes creating new tools and methods for processing text, images, and audio. A GAN is a class of machine learning framework and is the next emerging network in deep learning applications. GANs have the feasibility to build improved models, as they can generate the sample data as per application requirements. There are various applications of GAN in science and technology, including computer vision, security, ultimedia and advertisements, image generation and translation, text-to-images synthesis, video synthesis, high-resolution image generation, drug discovery, etc."– Provided by publisher.
Identifiers: LCCN 2022041650 (print) | LCCN 2022041651 (ebook) | ISBN 9781032068107 (hardback) |
ISBN 9781032068114 (paperback) | ISBN 9781003203964 (ebook)
Subjects: LCSH: Machine learning. | Neural networks (Computer science)
Classification: LCC Q325.5 .G44 2023 (print) | LCC Q325.5 (ebook) |
DDC 006.3/1–dc23/eng20221229
LC record available at https://lccn.loc.gov/2022041650
LC ebook record available at https://lccn.loc.gov/2022041651

ISBN: 9781032068107 (hbk)
ISBN: 9781032068114 (pbk)
ISBN: 9781003203964 (ebk)

DOI: 10.1201/9781003203964

Typeset in Palatino
by Newgen Publishing UK

# Contents

# Preface

A generative adversarial network (GAN) is a class of machine learning frameworks. In short, GANs are a methodology for generative modelling using deep learning techniques. GAN is a machine learning framework class and is the emerging network in deep learning applications. GANs are ingenious techniques for training a generative model by framing the problem as a supervised learning problem. Still, GANs also have a scope of research for semisupervised learning and reinforcement learning. GANs are a recent and rapidly changing field, which is distributing on the capability of generative models in their ability to generate genuine examples through a range of problem domains, most notably in image-to-image translation and video translation.

GANs have the feasibility to build improved models, as they can generate the sample data as per application requirements. There are various applications of GAN in science and technology, including computer vision, security, image editing, security, image generation, natural language processing, emotion detection, image translation, text-to-image synthesis, and video synthesis.

The main objective of this book is to focus on trending research in generative adversarial networks and deep learning, which comprises the design and development of innovative technologies and techniques for images, texts, and audio. The main objective of this book is to provide insights on deep learning generative models not only for images but also for text and audio.

This book comprises 14 chapters. It was impossible to include all current aspects of the research in the targeted areas. The book, however, presents a valuable tool for several different methodologies to be applied for various applications using GAN. Each chapter reflects a variety of application fields and methods.

Chapter 1 presents generative adversarial networks techniques and their use cases. It explains the difference between autoencoders and GAN, and the difference between VAN and GAN. The applications cover various applications of GAN models, which includes 3D object production used in medicine, GAN for image processing, face detection, and texture transfer, traffic control.

Chapter 2 focuses on image-to-image translation using GAN. The process of translating one image into another form/style according to the user's requirements is known as image-to-image (I2I) translation. It is a component of computer vision application that aims to understand the relationship between an input image and an output image and map them while retaining the original image's fundamental properties or attributes. Many computer vision-related tasks are used in image translation, including image generation, pose estimation, and image segmentation, high-resolution image. There are a variety of approaches to translating images. Generative modelling is a widely used technique for image-to-image translation. This process is being carried out using GAN. Because they are faster and learn the distributions more extensively, networks outperform other machine learning techniques. This chapter focuses on image-to-image translation strategies, focusing on style transmission. Deep convolutional GAN, CycleGAN, ConditionalGAN, and StarGAN are some instances of GAN.

Chapter 3 analyses the role of GAN in image editing. GAN is a neural network architecture for generative modelling. Generative modelling generates new examples that plausibly come from an existing distribution of samples, such as developing similar but different images from a dataset of existing ideas. Image editing is changing the content of an image to create a new image that meets the user's requirements and expectations. A GAN is made up of a generator and a discriminator positioned against each other. The discriminator tries to distinguish between the created image and the actual image from the database. At the same time, the generator is assigned to produce images similar to those in the database. This unusual interaction eventually trains the GAN and fools the discriminator into mistaking the generated images for images from the database.

Chapter 4 analyses the GAN for video-to-video translation. It explains how to use video-to-video interpretation to solve various problems, such as video super-goal, video colourization, and video division. For video-to-video interpretation, a common approach is to use a nearby edge-shrewd worldly consistency to fuse the transient misfortune between nearby casings in the improvement. Instead, we provide a blunder-based instrument that ensures the video-level consistency of the same area in diverse layouts. To get more consistent recordings, both global and local consistency in our video-to-video framework is combined simultaneously.

Chapter 5 presents the security issues in GAN. Security and its accompanying charge are an iterative pair of items that evolve in response to one another's advancements. The goal of this research is to look at the various ways in which GANs have been used to provide security improvements and attack scenarios to get beyond detection systems. This research aims to examine recent work in GANs, particularly in the areas of device and network security. It also discusses new challenges for intrusion detection systems based on GANs. In addition, it explores several potential GAN privacy and security applications and some future research prospects.

Chapter 6 presents the three major areas of design of intrusion detection systems where GANs are used. It explains the GANs to handle the problem of an imbalanced data set, in which the minority class samples are oversampled using a GAN generator to mitigate the imbalance. Next, it covers some recent research that focuses on employing GANs as anomaly detectors by using the reconstruction and discrimination loss as anomaly scores. When fed a malign class data sample, GAN's anomaly detection framework learns the distribution of benign class data samples, resulting in a more extensive reconstruction and discrimination loss. Finally, it addressed some well-known outcomes in adversarial machine learning, in which GANs are trained to create adversarial examples against IDS based on machine learning/deep learning.

Chapter 7 covers the case study of textural description to facial image generation. Text-to-image creates a realistic image that corresponds to a text description. The purpose of facial image generation from text is to generate the suspect's face from the victims' descriptions. Machine learning and deep learning techniques were used to extrapolate specific use cases from the generic training data provided to the machine learning architecture. Convolutional neural networks (CNNs) is a type of deep learning architecture that has been employed in the past. On the other hand, these architectures are best suited for complicated visual tasks like image segmentation and recognition. This study effort has been proposed using GANs to construct a suspect's face. GAN's unique feature is its capacity to undertake indirect training. A generator is trained on the outputs of a discriminator that is also being trained simultaneously. GAN allows to produce vivid facial images with a variety of text descriptions. A generator makes realistic-looking images based on the training, while the discriminator learns to distinguish between actual and phony images. This chapter proposes that the text be encoded into vectors using Universal Sentence Encoder, which can then be utilized for text categorization. Deep convolutional generative adversarial networks (DCGANs) is used for the image generating process.

Chapter 8 presents the application of GAN in natural language generation. Natural language generation involves the automatic synthesis of text, reports, or dialogues that can be included in artificial intelligence (AI)-powered automated devices. With the use of generative adversarial networks, it is now possible to develop a natural language generation network. The use of GAN in natural language generation is discussed in this chapter, along with problems, approaches, and applications.

Chapter 9 covers how GAN is used for audio synthesis along with its application. GANs are a sophisticated way for creating remarkable images. However, researchers have struggled to apply them to more sequential data, such as audio and music, where autoregressive (AR) models, such as wave nets and transformers, prevail by predicting an odd pattern at a time. While this aspect of AR styles helps to their success, it also has the capability to the extent that the example is excruciatingly sequential and slow, requiring solutions like spreading and special kernels for a real-time generation.

Chapter 10 presents a study on the application domains of electroencephalogram for deep learning-based transformative health care. Electroencephalography (EEG) uses metal discs called electrodes connected to the human scalp to measure currents that flow within our brain during task performance. It's being used in a variety of healthcare-related fields, including game influence analysis on mental health, rehabilitative help prescription, physical and mental health monitoring, and so on. Deep learning, on the other hand, is a technique that is widely employed in the medical field. This chapter describes various electroencephalogram (EEG) electrodes and their specific applications in various domains of deep learning-based revolutionary healthcare. The existing literature in the related domain has been analysed based on several benefits and performance flaws. The gaps discovered in the same are used to conduct an extensive study in this domain to find a possible research direction for future researchers.

Chapter 11 presents the application of GAN for emotion detection. Different GAN architectures are used in the research to construct game-based apps, improve their quality, balance difficulty levels, customise characters, and study the game's positive and negative effects on the players' mental health. In the sequestered life during pandemics, the popularity of video games such as real-time strategy games, arcade games, first-person shooter games, text adventure games, team sports games, and racing games has also increased among young people. This chapter examines recent work in the field and its benefits and downsides. Deep neural networks and GAN-based frameworks have had their working mechanisms explored in gaming applications. Finally, along with the concluding statements, some possible directions have been presented as a roadmap for future researchers.

Chapter 12 presents the GAN application for the enhancement of underwater images. The technique described in this chapter addresses some of the issues in underwater imaging. Deep learning technology's use has increased significantly due to its ability to tackle complex problems and discover hidden patterns in the data. GAN, which consists of a generator network and a discriminator network, is used to enhance image translation using underwater images (corrupted). On the UNET architecture, a generator network is created with the addition of fully connected convolution. A novel hues colour loss objective function is introduced to train this GAN architecture, the influence of which can be seen in the results as the suggested technique outperforms many existing approaches.

Chapter 13 presents the GAN challenges and its optimal solutions. The applications of GANs have received massive recognition in the field of artificial intelligence, opening up a wide variety of exciting research opportunities. Though GANs have gained widespread adoption, they face a number of obstacles when it comes to training the model.

This chapter goes through certain difficulties that might arise when training two neural networks in GAN, such as reaching Nash equilibrium, vanishing gradient problems, mode collapse, and non-convergence issues. The chapter explains the experiment that was carried out to demonstrate vanishing gradient issues. Following that, it addressed the various researchers' proposed solutions to the issues that arise during the training of a GAN model. This chapter also discussed the future directions for finding the improved training methods and GAN model versions.

**Dr. Roshani Raut**
*Pimpri Chinchwad College of Engineering, Savitribai Phule Pune University, Pune, India*
**Dr. Pranav Pathak**
*MIT School of Bioengineering Sciences and Research, Pune, India*
**Dr. Sachin Sakhare**
*Vishwakarma Institute of Information Technology, Pune, India*
**Dr. Sonali Patil**
*Pimpri Chinchwad College of Engineering, Savitribai Phule Pune University, Pune, India*

# *Editors*

**Roshani Raut** has obtained Ph.D. in Computer Science and Engineering and is currently working as an Associate Professor in the Department of Information Technology and Associate Dean, International Relations, at Pimpri Chinchwad College of Engineering, Savitribai Phule Pune University, Pune, Maharashtra, India. She has received various awards at the national and international level for research, teaching, and administration work. She worked an author/editor for various upcoming books of IGI Global, CRC Taylor and Francis, and Scirvener Wiley. She worked on various conference committees as a convener, TPC member, and Board member. She has availed the research and workshop grants from BCUD, Pune University. She has presented more than 70 research communications in national and international conferences and journals. She has 15 patents to her name, out of which three patents have been granted. Her area of research is artificial intelligence, machine learning, data mining, deep learning, and Internet of Things. For more details: https://roshaniraut531.wixsite.com/roshaniraut

**Pranav D Pathak** is Associate Professor in MIT School of Bioengineering Sciences and Research, Pune. He received his undergraduate, postgraduate degrees from Amravati University and doctoral degrees from Visvesvaraya National Institute of Technology, Nagpur, India. His research interests include use of Indian traditional knowledge in medicines, biorefinery, adsorptive separation, solid waste utilization, and wastewater treatment. He has published more than 25 articles in SCI/SCIE journals and book chapters. He has attended more than 30 national/international conferences. Currently he is the editorial board member and reviewer of a few international journals.

**Sachin R Sakhare** is working as Professor in the Department of Computer Engineering of the Vishwakarma Institute of Information Technology, Pune, India. He has 26 years of experience in engineering education.

He is recognized as Ph.D. guide by Savitribai Phule Pune University and currently guiding seven Ph.D. scholars. He is a life member of CSI, ISTE, and IAEngg. He has published 39 research communications in national, international journals and conferences, with around 248 citations and with an H-index of 6. He has authored six books. He worked as a reviewer of journals published by Elsevier, Hindawi, Inderscience, IETE, and Springer. He worked as a reviewer for various conferences organized by IEEE, Springer, and ACM. He has also worked as a member of the technical and advisory committees for various international conferences. Dr. Sachin has delivered invited talks at various conferences, Faculty Development Programs (FDPs), and Short Term Training Programs (STTPs) as well as to postgraduate and Ph.D. students. He has guided 26 postgraduate students. Two Indian and two Australian patents have been granted in his name.

Dr. Sachin is associated with international universities also. He is an invited guest faculty at Amsterdam University of Applied Sciences (AUAS), Netherlands, to conduct sessions on data structures and Internet of Things. He has also visited Aalborg University, Copenhagen, Denmark and RWTH Aachen University, Germany, for academic collaboration.

He has availed the research workshop grants from AICTE, BCUD, Pune University. He has filed and published four Indian patents and six copyrights. One Australian patent has been granted in his name. He has delivered a session on Outcome Based Education and NBA accreditation. He has completed the course on ABET accreditation.

**Sonali Patil** (IEEE Senior Member) is a prominent educationalist currently working as a Professor and Head of the Information Technology Department, Pimpri Chinchwad College of Engineering (PCCoE), Savitribai Phule Pune University (SPPU), Pune, Maharashtra, India. She has been actively engaged in academics for the last 19+ years. Her qualifications include B.E. (computer science and engineering), Masters in computer engineering, and Ph.D. in computer engineering.She has published more than 70 research papers in reputed international journals and conferences to date in addition to holding international and national patent publications. She is conferred with the "IEEE Innovator/Researcher of The Year Award-2021" by IEEE Pune section and "Women Researcher Award" by Science Father Organization, India. She has delivered expert talks on multiple platforms and continues to avidly contribute her pedagogical expertise to the field of IT.

# *Contributors*

**Aarti Amod Agarkar**
MMCOE, Pune, India

**Jayashri Bagade**
Vishwakarma Institute of Information Technology, Pune, India

**Parul Bhanarkar**
Jhulelal Institute of Technology, Nagpur, India

**Gajanan K. Birajdar**
Ramrao Adik Institute of Technology, Mumbai, India

**Pradnya Borkar**
Jhulelal Institute of Technology, Nagpur, India

**Yogini Borole**
Savitribai Phule Pune University, Pune, India

**Yogesh Dandawate**
Vishwakarma Institute of Information Technology, Pune, India

**Sima Das**
Maulana Abul Kalam Azad University of Technology, West Bengal, India

**Digvijay Desai**
Vishwakarma Institute of Information Technology, Pune, India

**Nisha Singh Gaur**
Ramrao Adik Institute of Technology, Mumbai, India

**Ahona Ghosh**
Maulana Abul Kalam Azad University of Technology, West Bengal, India

**Yash Goda**
Dwarkadas J. Sanghvi College of Engineering, Mumbai, India

**Anuja Jadhav**
Pimpri Chinchwad College of Engineering, Pune, India

**Swati Jaiswal**
Pimpri Chinchwad College of Engineering, Pune, India

**Abhishek Kasar**
Vishwakarma Institute of Information Technonlogy, Pune, India

**Atul B. Kathole**
Pimpri Chinchwad College of Engineering, Pune, India

**Vatsal Khandor**
Dwarkadas J. Sanghvi College of Engineering, Mumbai, India

**Harmeet Kaur Khanuja**
MMCOE, Pune, India

**V. Kumar**
THDC-IHET, Tehri, India

**Ramchandra Mangrulkar**
Dwarkadas J. Sanghvi College of Engineering, Mumbai, India

**Hirkani Padwad**
Shri Ramdeobaba College of Engineering and Management, Nagpur, India

**Sonali D. Patil**
Pimpri Chinchwad College of Engineering, Pune, India

**Mukesh D. Patil**
Ramrao Adik Institute of Technology, Mumbai, India

**Suchitra Paul**
Brainware University, Kolkata, India

**Naitik Rathod**
Dwarkadas J. Sanghvi College of Engineering, Mumbai, India

**Roshani Raut**
Pimpri Chinchwad College of Engineering, Pune, India

**Nemil Shah**
Dwarkadas J. Sanghvi College of Engineering, Mumbai, India

**Chaitrali Sorde**
Pimpri Chinchwad College of Engineering, Pune, India

**Reena Thakur**
Jhulelal Institute of Technology, Nagpur, India

**Kapil N. Vhatkar**
Pimpri Chinchwad College of Engineering, Pune, India

**Shreyash Zanjal**
Vishwakarma Institute of Information Technonlogy, Pune, India

# 1

# *Generative Adversarial Networks and Its Use Cases*

**Chaitrali Sorde, Anuja Jadhav, Swati Jaiswal, Hirkani Padwad, and Roshani Raut**

**CONTENTS**



## 1.1 Introduction

Generative adversarial networks (GANs) have recently been a popular study area. "GANs are the most exciting notion in machine learning in the last 10 years," claimed Yann LeCun, a deep learning legend, in a Quora post. According to Google Scholar, there are a vast number of papers relevant to GANs. In 2018, there are approximately 11,800 papers

DOI: 10.1201/9781003203964-1

connected to GANs as well as 32 papers per day and more than one paper every hour linked to it. A generator and a discriminator are the two models that make up a GAN.

Predictability minimization [1,2] is the most relevant work. Emami et al. [3,4] discuss the relationship between predictability minimization and GANs. GANs have been the subject of multiple previous reviews due to their popularity and relevance. The following is a summary of the differences between this and earlier studies. GANs for specific applications: There have been surveys of GANs being used for specific applications such as picture synthesis and editing [5], audio improvement and synthesis, and so on.

A GAN comprises of two contending neural organizations: a generative organization known as a generator and assigned G, and a discriminative organization known as a discriminator and stamped D, the two of which are enlivened by the two-player minmax game. The generator network attempts to deceive the discriminator by making sensible examples, while the discriminator attempts to recognize genuine and counterfeit examples. A strategy can be utilized as the generator and discriminator organizations, as long as the generator can become familiar with the circulation of the preparation information and the discriminator can extricate the component to order the result.

A convolutional neural network or a recurrent neural network can be used as the discriminator network, while a de-convolutional neural network can be used as the generator network. As a result, GANs can be used to build multidimensional data distributions like pictures. GANs have shown potential in a variety of difficult generative tasks, including text-to-photo translation, picture generation, image composition, and image-to-image translation [3]. GANs are a powerful type of deep generative model; however, they have a variety of training issues, such as mode collapse and training instability. There are different types of learning approaches in machine learning such as supervised and unsupervised learning.

## 1.2 Supervised Learning

The PC is educated as a visual cue in administered learning. It takes what it has gained from past information and applies it to current information to estimate future occasions. Both the info and designated yield information are helpful in foreseeing future occasions in this situation.

The input data is labeled or tagged as the correct response for accurate predictions. Figure 1.1 shows the supervised learning approach.

Regulated learning, similar to all artificial intelligence (AI) calculations, depends on preparing. In Figure 1.1. the framework is provided named informational indexes all through its preparation stage, which helps know the yield related to every particular information esteem. The prepared model is then given test information, which is marked information with the names stowed away from the calculation. The testing information is utilized to decide how well the calculation performs on unlabeled information.

By consistently estimating the model's results and tweaking the framework to draw nearer to its objective precision, the regulated educational experience is moved along. The degree of exactness still up in the air by two factors: the marked information accessible and the calculation used.

Information from preparing should be adjusted and scrubbed. Information researchers should be cautious with the information the model is prepared on the grounds that trash or

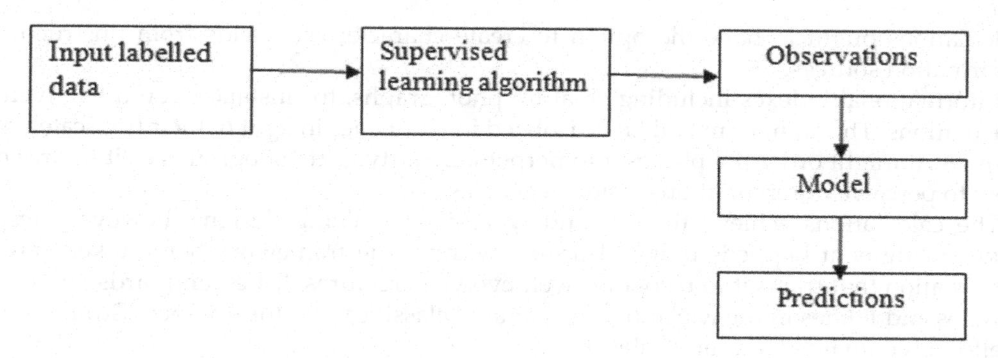


**FIGURE 1.1**
Supervised learning approach.

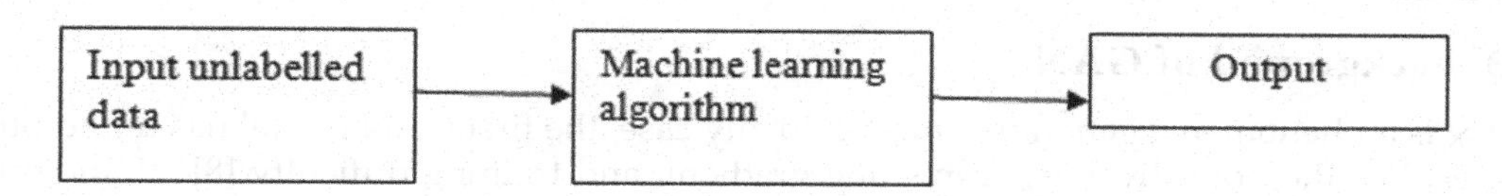


**FIGURE 1.2**
Unsupervised learning approach.

copy information will slant the AI's perception. The variety of the information influences how really the AI performs when confronted with new situations; assuming that the preparation information assortment contains inadequate examples, the model will vacillate and neglect to give dependable reactions. Order and relapse are the two fundamental results of administered learning frameworks.

### 1.2.1 Unsupervised Learning

Unsupervised learning, on the other hand, is a technique for teaching computers to use data that has not been classified or labeled. It means that no preparation information is accessible and the machine is customized to learn all alone. With no earlier information on the information, the machine should have the option to group it.

The objective is to open the machines to huge measures of different information and allow them to gain from it to uncover already obscure bits of knowledge and reveal stowed away examples. Accordingly, solo learning calculations don't necessarily create unsurprising outcomes. Rather, it figures out what makes the given dataset novel or interesting.

It is important to program the machine to learn all alone. Both organized and unstructured information should be perceived and examined by the PC. Solo learning calculations can deal with more complicated handling undertakings than regulated learning frameworks. Besides, one method of scrutinizing AI is to utilize solo learning. Here's an accurate illustration of unsupervised learning in Figure 1.2.

The objective of solo learning is for calculations to identify designs in preparing informational collections and order input things in view of those examples. To separate applicable data or highlights, the calculations dissect the informational collections' central construction. Accordingly, by searching for joins between each example and info thing, these calculations ought to have the option to create characterized yields from unstructured information sources.

Informational indexes including creature photographs, for instance, could be given to calculations. The animals may then be isolated into bunches in light of their fur, scales, and plumes. It might order the photos into perpetually nitty gritty subgroups as it figures out how to perceive differentiations inside every class.

The calculations achieve this by finding and recognizing designs; however, design acknowledgment happens in solo learning without the framework being taken care of information that trains it to recognize well-evolved creatures, fishes, and birds, or among canines and felines in the warm-blooded animal classification, for instance. Grouping is a well-known unaided learning subject [7].

## 1.3 Background of GAN

GAN is a phenomenal generative model. In any case, the first GAN model has numerous issues, like the poor diversity, vanishing gradient, and training difficulty [8]. Numerous endeavors have been committed to getting better GANs through various enhancement strategies. Hence, beginning around 2014, speculations and articles connected with GAN have arisen in an interminable stream, and numerous new GAN-based models have been proposed to work on the security and nature of the produced outcomes.

A generative adversarial network (GAN) is a sort of AI proposed by Ian Goodfellow and his associates in a course named "Generative Adversarial Nets" [6] in June 2014. In a game, two brain networks go up against each other. This cycle decides how to produce new data with comparable experiences as the planning set given an arrangement set. For instance, a GAN prepared on pictures can create new photographs that are cursorily real to human spectators and have an assortment of helpful properties. GANs, which were first introduced as a kind of generative model for independent learning, have likewise demonstrated helpful for semi-directed learning, support learning, and completely administered learning.

The "backhanded" prepping through the discriminator, which is also being updated gradually, is at the heart of a GAN's central concept. This simply means that the generator is attempting to deceive the discriminator rather than limiting the distance to a certain image. This allows the model to learn without being supervised.

Setting two calculations in opposition to one another started with Arthur Samuel, a noticeable scientist in the field of software engineering who's credited with promoted the expression "machine learning." While at IBM, he concocted a checkers game—the Samuel Checkers-playing Program—that was among the first to effectively self-learn, to some degree by assessing the shot at each side's triumph at a given position.

In 2014, Goodfellow and his associates provided the main working execution of a generative model dependent on adversarial networks.

Goodfellow has frequently expressed that he was roused by noise contrastive assessment, a method of learning information dissemination by looking at it against a noise distribution, that is, numerical capacity addressing distorted or corrupted information. The noise contrastive assessment involves a similar loss function as GANs—at the end of the day, a similar proportion of execution concerning a model's capacity to expect anticipated results.

Once more, GANs comprise of two sections: generators and discriminators. The generator model produces engineered models (e.g., images) from arbitrary noise sampled utilizing a distribution model, which alongside genuine models from a preparation informational collection are taken care of to the discriminator, which endeavors to recognize the two. Both the generator and discriminator work on in their particular capacities until the discriminator can't perceive the genuine models from the integrated models with better than the half precision expected of possibility.

GANs are algorithmic plans that utilize two brain affiliations, setting one as opposed to the accompanying to make new, created cases of information that can be pass with genuine information. They are utilized regularly in picture age, video age and voice age.

Generative adversarial networks embraces up a game-hypothetical strategy, rather than standard brain organizations. The organizations advance out a viable method for making from a preparation appropriation through a two-player game. The two enemies are generator and discriminator. These two foes are in reliable fight all through the arranging correspondence. Since an ill-disposed learning technique is taken on, we genuinely need not care about approximating obstinate thickness capacities.

One brain organization, called the generator, delivers new data events, while the other, the discriminator, evaluates them for validness; for instance, the discriminator closes whether each instance of data that it reviews has a spot with the certifiable planning dataset or not.

As you can recognize from their names, a generator is used to make genuine looking pictures and the discriminator's liability is to perceive which one is phony. The components/enemies are in a consistent battle as one (generator) endeavors to deceive the other (discriminator), while various takes the necessary steps not to be deceived. To make the best pictures, you will require a marvelous generator and a discriminator. This is since, in such a case that your generator isn't sufficient, it will not at any point have the choice to deceive the discriminator and the model will not at any point join together. Accepting the discriminator is terrible, then, picture which has no perceptible pattern will similarly be named authentic and in this manner your model never plans and consequently you never convey the best outcome. The information and unpredictable upheaval can be a Gaussian appointment and characteristics can be inspected from this course and dealt with into the generator association and a picture is created. This made the picture differentiated and a genuine picture by the discriminator and it endeavors to perceive if the given picture is phony or certified.

### 1.3.1 Image-to-Image Translation

GAN-based methods have been extensively used in picture to-picture interpretation and conveyed drawing in results. Inpix2pix, contingent GAN (cGAN), was used to get to know a preparation from an info picture to a result picture; cGAN gains a prohibitive generative model using matched pictures from source and target regions. CycleGAN was utilized for picture to-picture interpretation endeavors in the nonparticipation of joined models. It takes in a preparation from a source region A to a goal space B by introducing two cycle-consistency disasters. Likewise, DiscoGAN and DualGAN use an independent learning approach for picture-to-picture interpretation reliant upon unaided data, yet with different misfortune capacities. Consonant GAN utilizes for solo picture to-picture understanding presents spatial smoothing to maintain unsurprising mappings during translation.

Image-to-image interpretation is to take in a preparation between pictures from a source region and pictures from an objective region and has various applications including picture colorization, creating semantic names from pictures [9], supergoal pictures, and space variety. Many picture-to-picture interpretation approaches require regulated learning settings in which sets of looking at source likewise target pictures are open. In any case, acquiring consolidated planning data is expensive or a portion of the time shocking for grouped applications.

Overall, a picture-to-picture interpretation method requirements to recognize spaces of interest in the information pictures and sort out some way to make a translation of the recognized locales into the objective region. In an unaided setting with no paired pictures between the two spaces, one ought to zero in on the spaces of the picture that are at risk to move. The task of finding spaces of interest is more critical in uses of picture-to-picture interpretation where the understanding should be applied extraordinarily to a particular kind of article as opposed to the whole picture.

CycleGANs empower taking in a planning starting with one area A then onto the next space B without finding impeccably coordinated, preparing sets. Assume we have a bunch of images from the area A and an unpaired arrangement of images from the space B. We need to have the option to decipher an image starting with one set then onto the next. To do this, we characterize a planning G(G: A→B) that makes an honest effort to plan A to B. Be that as it may, with unpaired information, we presently don't can take a gander at genuine and counterfeit sets of information. Yet, we realize that we can change our model to create a result that has a place with an objective space.

So when you push an image to domain A, we can prepare a generator to deliver reasonable looking images to domain B. In any case, the issue with that will be that we can't constrain the result of the generator to relate to its input. This prompts an issue called mode breakdown in which a model may plan different contributions from area A into a similar result from space B. In such cases, given an info to domain B, all we know is that the result should resemble domain B. In any case, to get the right planning of the contribution to the relating objective space we present an extra planning as reverse planning G′(G′: B→A) which attempts to plan B to A. This is called cycle-consistency imperative.

Assuming if we decipher an image from domain A to an image of domain B, and afterward we interpret back from a domain B to a domain A, we ought to show up back at a similar image of the pony with which we began.

A total interpretation cycle ought to carry you back a similar image with which you began with. On account of image translation structure space A to B, assuming the accompanying condition is met, we say that the change of an image from area A to area B was right.

$$G_{AtoB}(G_{AtoB}(x)) \approx x \tag{1.1}$$

CycleGAN ensures that the model takes in a right planning from space A to area B. The steps required to perform image-to-image translation requires following steps:

1. Visualization of datasets
2. Defining an appropriate model
3. Make use of discriminator
4. Use of residual blocks/functions
5. Make use of generator
6. Training process
   a. Calculation of discriminator and generator function
   b. Make use of optimizer to get proper results

## 1.4 Difference between Auto Encoders and Generative Adversarial Networks

### 1.4.1 Auto Encoders

An auto encoder's job is to learn both the encoded and decoded networks simultaneously. To put it simply, the encoder receives an input (such as an image) and works hard to compress it so that it may be transmitted to the decoder in a highly compressed state.

This encoding/decoding is learned by the network since the loss measure grows as the discrepancy between the input and output images grow. Decoding and encoding increase with each cycle, as the encoder improves its capacity to discover an efficient compressed form of the input information.

### 1.4.2 Generative Adversarial Networks

Here, we've got a "generator" whose job is to transform a noise signal into a target space (again, images are a popular example). It is the "discriminator" of the second component (the adversary) whose task it is to discriminate between true and fraudulent images drawn from the target space. There are two unique stages of training in this example, each with a different loss:

For the first time, the discriminator is presented with tagged examples taken from the real collection of photos and fabricated by the generator (of course, at the start, these images are just noise). A binary classification loss measure is used to calculate the loss (e.g. cross entropy). Thus, the discriminator learns to discriminate between a false and an accurate image, as a result of this training.

However, before the generator has learned too much, we must get to it. The discriminator is trained by increasing the generator's noise and then validating the generator's output. Finally, we want the discriminator (the machine learning system) to recognize the generated image as real, which implies that the generator has learned how to build an image that resembles a "real" image from the training set, and the discriminator has been deceived.

## 1.5 Difference between VAN and Generative Adversarial Networks

The auto encoder is another network architecture that is used to encode object, such as images into latent variables. The latent variables usually have far less dimension and less parameter than the original object. We usually only use the encoder part after we're finished with the training with auto encoder. Another use of encoder part of autoencoder is that it can used to initialize a supervised model. Usually fine-tune the encoder jointly with the classifier.

The autoencoder is usually comprised of these modules shown in Figure 1.3.

However, simply using the autoencoder to generate images will not be considered as a generative model. Say that you input an image, then the output image you generated will always be the same. In another word, the decoder part of autoencoder network is simply generating things it already remembered.

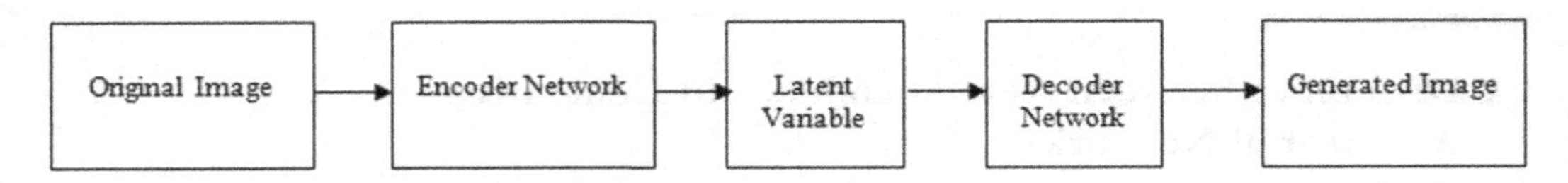


**FIGURE 1.3**
Architecture of auto encoder.

The VAE addresses the abovementioned problem by supplying the latent variables with a sampling technique that makes each feature unit Gaussian or some other distribution. In other words, we'll use the "sampled latent vector" instead of the true latent vector. In the next section, we will discuss these deductions in detail.

## 1.6 Application of GANs

Let's look at how and where GANs are being used in the real world instead of focusing on theory and academic and nonacademic research. When it comes to putting those theories into practice, it appears that there has only been a slow start, and there are good reasons for why. Photorealistic graphics may already be generated with GANs in a variety of industrial and interior design applications as well as in the gaming industry. It has also been reported that GANs have been used to train members of the film and animation industries. A three-dimensional representation of an object can be reconstructed from fragmentary images collected during astronomical observation.

### 1.6.1 Application of GANs in Healthcare

Because of the potential for picture improvement, GANs in medicine can be used for photorealistic single image super-resolution. The growing need for GANs in healthcare is a direct result of the more stringent requirements for images used in medical imaging. Certain measurement procedures can make it challenging to get high-quality images. There is a considerable need to limit the radiation effect on patients when employing low-dose scanning in computer tomography (CT, to reduce the negative effect on persons with certain health preconditions such as lung cancer) or magnetic resonance imaging (MRI). It is more difficult to get high-quality photos because of the poor quality scans.

Even though GAN deployment in the medical area is a slow process due to the multiple experiments and trials required due to safety concerns, super-resolution improves gathered images and efficiently reduces noise. Care must be taken to guarantee that the denoising does not misrepresent the image's genuine content, which could lead to a misdiagnosis, by enlisting the assistance of several domain professionals.

Advertising and marketing are the industries that are most likely to use GAN. When it comes to marketing products and services, it is common to be faced with the need to provide both original and repetitive content. In order to take use of this new technology, Rosebud AI has developed a Generative Photos app. The software creates images of fictitious models. This is performed by inserting fictitious faces into stock photos of actual models. Replace the default face with a custom one and apply various edits to it; this is where things get very interesting.

### 1.6.2 Applications of Generative Models

Following are some applications of the GAN model used in generative environment.

#### *1.6.2.1 Generate Examples for Image Datasets*

GAN exhibits a reasonable capacity to make practical human face photos. They're so similar, as a matter of fact, that the end-product is staggering. Therefore, the discoveries got a great deal of press inclusion. Face ages have been shown about VIP ancestors, along these lines the created faces incorporate components of current hotshots, causing them to seem natural yet not exactly. [10].

#### *1.6.2.2 Generate Realistic Photographs*

GAN's technique aids in the creation of synthetic pictures. BigGAN images that are nearly indistinguishable as real-life photographs [11].

#### *1.6.2.3 Generate Cartoon Characters*

We can generate characters with GAN by taking human faces as input and processing them to get a high density polygon, which saves a lot of time when developing cartoon characters from the ground up [12].

#### *1.6.2.4 Image-to-Image Translation*

Photographs of cityscapes and buildings are created using semantic pictures. Satellite photos are translated by Google Maps. The images were changed from day to night. Black-and-white images are colored [13].

The following example shows four different photo translation scenarios:

- Creating a painting style from a photograph
- Zebra to horse translation
- Changing a summer photo into a winter one
- Creating a Google Maps presentation from a satellite image
- Making a photograph out of an artwork
- Creating a photograph from a sketch
- Translation from apples to oranges
- Turning a photograph into an artistic artwork.

#### *1.6.2.5 Text-to-Image Translation*

Given current state-of-the-art results, text-to-image synthesis is a difficult problem with a lot of space for improvement. Existing approaches provide synthesized images with a crude sketch of the stated image, but they fall short of capturing the genuine core of what the text says. To create realistic-looking photos from textual descriptions of simple objects such as birds and flowers.

#### 1.6.2.6 Semantic-Image-to-Photo Translation

Ting-Chun Wang et al. show how to utilize conditional GANs to create photorealistic images from a semantic image or sketch as input Photograph of a cityscape with a semantic image [10].

(a) Photograph of a bedroom, with a semantic image
(b) Photograph of a human face with a semantic image
(c) A photograph of a human face with a drawing.

They also show how to manipulate the resulting image using an interactive editor.

#### 1.6.2.7 Photos to Emojis

Yaniv Taigman and colleagues developed a GAN to convert images from one domain to another, such as street numbers to MNIST handwritten digits and celebrity shots to emoticons (little cartoon faces).

#### 1.6.2.8 Photograph Editing

Guim Perarnau and colleagues used a GAN, specifically their IcGAN, to reconstruct images of faces with specific characteristics like as hair color, style, facial expression, and even gender.

#### 1.6.2.9 Face Aging

Because the appearance of the face changes with age, face aging is the process of rendering a face in order to estimate its future look, which is important in the field of information forensics and security.

## 1.7 Conclusion

The overview contains new GANs research on different subjects reaching out from images steganography to prepare the framework to scan the images. Additionally, the work covers many distinct types of GANs and GAN variants that researchers have utilized to solve essential image to image segmentation. It discusses how GANs have improved image quality. The utilization of GAN instability is in its earliest stages, yet it will probably not remain as such for a long time with quality of work.

## References

[1] I. Goodfellow, J. Pouget-Abadie, M. Mirza, B. Xu, D. Warde-Farley, S. Ozair, A. Courville, and Y. Bengio, Proceedings of the International Conference on Neural Information Processing Systems (NIPS 2014), 2017, pp. 2672–2680.

[2] K. Wang, G. Chao, Y. Duan, Y. Lin, X. Zheng, and F.Y. Wang, "Generative adversarial networks: introduction and outlook," IEEE/CAA J Autom Sin., 2017, 4:588–598. doi: 10.1109/JAS.2017.7510583
[3] H. Emami, M. MoradiAliabadi, M. Dong, and R.B. Chinnam, "SPA-GAN: spatial attention GAN for image-to-image translation," *IEEE Transactions on Multimedia*, 2020, 23:391–401.
[4] E. Agustsson, M. Tschannen, F. Mentzer, R. Timofte, and L. Van Gool, "Generative adversarial networks for extreme learned image compression," in Proc. ICCV, 2019, pp. 221–231.
[5] D. Bau et al., "GAN dissection: Visualizing and understanding generative adversarial networks," arXiv preprint arXiv:1811.10597, 2018.
[6] K. Kurach, M. Luˇci´c, X. Zhai, M. Michalski, and S. Gelly, "A large-scale study on regularization and normalization in GANs," in Proc. Int. Conf. Mach. Learn., 2019, pp. 3581–3590.
[7] M.-Y. Liu, T. Breuel, and J. Kautz, "Unsupervised image-to-image translation networks," *Advances in neural information processing systems*, 2017, 30.
[8] A. Radford, L. Metz, and S. Chintala, "Unsupervised representation learning with deep convolutional generative adversarial networks," arXiv preprint arXiv:1511.06434, 2015.
[9] A. Radford, L. Metz, and S. Chintala, Unsupervised Representation Learning DCGAN, ICLR, 2016.
[10] I.P. Durugkar, I. Gemp, and S. Mahadevan, "Generative multiadversarial networks," CoRR, abs/1611.01673, 2016.
[11] I. Goodfellow, J. Pouget-Abadie, M. Mirza, B. Xu, D. WardeFarley, S. Ozair, and A. Courville and Y. Bengio, "Generative Adversarial Networks," no. arXiv:1406.2661v1, p. 9, 2014.
[12] Y. Kataoka, T. Matsubara, and K. Uehara, "Image generation using generative adversarial networks and attention mechanism," in IEEE/ACIS 15th International Conference on Computer and Information Science (ICIS), Okayama, 2016, pp. 1–6.
[13] P. Domingos, "A few useful things to know about machine learning," Communications of the ACM, 2012, 55.10:78–87.

# 2

# *Image-to-Image Translation Using Generative Adversarial Networks*

**Digvijay Desai, Shreyash Zanjal, Abhishek Kasar, Jayashri Bagade, and Yogesh Dandawate**

**CONTENTS**



DOI: 10.1201/9781003203964-2

## 2.1 Introduction

Human beings are distinctive in many ways. We observe the world around us, and even though we see the same imagery, our mind makes myriad creative representations out of it. These representations can manifest themselves in distinctive ways. One of the earliest creative pursuits of the mind, visual illustration or drawing, can be attributed to this process. A little kid can paint and draw pretty well. This ability to make a graphic representation of an object or an idea by lines and colours can be computationally mimicked! The image-to-image (I2I) translation systems are a promising step in this direction.

We can define the goal of the I2I translation system as learning the relationship between the source input and output image utilising a set of image pairings in order to translate/generate images. While seeing an image and learning representation is an important task, the ways these representations can be used has very interesting and wide-reaching implications. This may include creating a photo with all the filled pixels from a simple sketch, translating a landscape image in winter to summer, and even making synthetic data for training neural networks. We can classify the I2I methods into two categories: two-domain and multi-domain I2I. The two-domain I2I deals with input and output images from two discrete domains. Since images do not change abruptly, no discernible gap exists between different domains, so multi-domain I2I algorithms exist.

Section 2.2 delves into the traditional I2I translation methods. Many hand-engineered techniques were popular prior to the development of deep learning approaches. These are briefly discussed, as well as the limitations of these methods. In recent years, advancements in adversarial machine learning and particularly the development of generative adversarial networks [3] have helped the I2I tasks tremendously. The I2I task can easily be formulated as a min-max algorithm. Image-based generative adversarial networks like deep convolutional GAN (DCGAN) [5] are excellent methods for a variety of I2I tasks. A lot of recent progress in GANs is geared toward I2I efforts. Section 1.3 takes a closer look at generative adversarial networks.

We can divide the I2I into two categories based on the images and their representations: multi-domain I2I and two-domain I2I. The two-domain I2I deals with images from two discrete domains as input and output. Because images do not change abruptly, there is no discernible gap between domains, so multi-domain I2I algorithms exist. According to the standard deep learning nomenclature, two-domain I2I methods can be divided into four broad categories.

1. *Supervised I2I*: Input and output images are provided for the supervised I2I algorithms. They want to find a function that can approximate this behaviour between the source and target classes and produce an image with the desired properties given the input image.
2. Unsupervised I2I: Supervised I2I is impractical due to the high cost of acquiring source and target image pairs. Unsupervised I2I algorithms attempt to predict the

distribution given two non-matched large sets of input domains and interconvert them using the underlying representations.
3. Semi-Supervised I2I: Semi-supervised algorithms use labelled and unlabelled data to introduce consistency and use mathematical techniques to do so. When compared to unsupervised algorithms, this produces more homogeneous results.
4. Few-shot I2I: The human learning approach is used to inspire few-shot learning algorithms. In comparison to the previous three categories, they use far fewer learning examples. This can be accomplished by investigating the data's similarity, structure, and variability, as well as the learning procedure's constraints.

Sections 2.4 to 2.7 go into great detail about these approaches and point out their flaws. The applications of these approaches are summarised, along with some illustrative examples. We discuss some additional work in two major categories, supervised and unsupervised I2I, in addition to popular and cutting-edge techniques.

## 2.2 Conventional I2I Translations

Imagine you take a picture of a person and create a drawing out of it. Is it possible to do it with the help of a computer? Well, the answer is yes. This type of image conversion falls into the domain of I2I translation. This technique aims to convert the input image $X_i$ from style $I$ to style $J$ to convert the image while retaining the basic properties of the original image. For example, the original photo($x_i$) you took is of style $I$, which needs to be translated to style $J$ with the help of a reference image $X_j$. Hence, we need to train the mapping model $M_{i\to j}$, which generates an image $X_{ij} \in J$ using reference image $X_j \in J$ and the input image $X_i \in I$.

Mathematically,

$$X_{ij} \in J : x_{ij} = M_{i\to j}\left(x_i\right) \tag{2.1}$$

This basic equation can be used for many applications in image processing and computer vision.

Image translation has been studied for decades. Different approaches have been exploited, including

- Filtering-based
- Optimisation-based
- Dictionary learning-based
- Deep learning-based
- GAN-based

### 2.2.1 Filtering-based I2I

Image convolution, a general-purpose filter effect for images, is used to implement filtering-based I2I. Convolution is a technique for transforming images by applying a kernel (filter) to each pixel and its local neighbours across the entire image. The kernel is a value matrix whose size and values affect the convolution process' transformation impact.

### 2.2.2 Optimisation-based I2I

The goal of employing optimisation techniques is to fine-tune particular parameters that affect profit and expenses so that the function can be maximised or minimised. Traditional methods, in general, fail to solve such large-scale issues, especially when nonlinear objective functions are involved. To tackle linear, nonlinear, differential, and non-differential optimisation problems, current techniques such as convex optimisation, defect detection, firefly optimisation, cuckoo optimisation, particle swarm optimisation, and others are utilised.

### 2.2.3 Dictionary Learning-based I2I

The goal of dictionary learning is to discover a frame (called a dictionary) in which some training data can be represented sparsely. The dictionary is better when the representation is sparse. The dictionary learning framework, which involves a linear decomposition of an input signal using a few fundamental elements learned from the data, has produced state-of-the-art results in a variety of image and video processing tasks. A dictionary is made up of these elements, which are known as atoms.

### 2.2.4 Deep learning-based I2I

When it comes to I2I translation, deep neural networks (DNNs) such as encoder–decoder Networks (e.g. [54]) is a remarkable invention. Encoder–decoder networks are a type of DNN with a specific structure that can convert one picture into another. Multiple down-sampling layers with increasing numbers of channels (the encoder part) are followed by varied upsampling layers with decreasing numbers of channels in an encoder–decoder network (the decoder part).

### 2.2.5 GAN-based I2I

In 2014, Ian Goodfellow et al. [3] developed a new deep neural network known as generative adversarial network (GAN). In this class of neural networks, two models fight (where one wins and the other loses and improves themselves in the process). It is a very popular I2I translation method thanks to its extraordinary capability in generating crisp, sharp images. The model learns to create fresh data that has the same statistics as the training data set. This newly generated data is as good as real-world data. Following the development of GANs, various researchers implemented different methods to create new types of GANs with the same base structure and a few add-ons. DCGAN, proposed by Alec Radford [5], is the same as a conventional GAN but with a more powerful convolutional neural network architecture. They were the first to successfully incorporate a convolutional neural network (CNN) directly into a full-scale GAN model. CycleGAN [4], proposed by Jun-Yan Zhu, is a type of GAN used for unpaired image-to-image (I2I) translation. It is usually used for transferring the characteristics of one image to another by reconstructing the original image. This chapter will be exploring such networks which are used for I2I translation. It will explore the architecture, applications, and limitations of these models.

## 2.3 Generative Adversarial Networks (GAN)

Let us discuss GANs in detail in this section. Generative adversarial network (GAN) is multi-layer perceptron (MLP)/CNN-based architecture developed by Ian Goodfellow [3] in 2014. It's a machine learning (ML) paradigm in which two neural networks compete against one other to improve their prediction accuracy. GANs are usually unsupervised and learn using a cooperative zero-sum game framework. GANs have very high potential as they can mimic any type of real image. It can also be used to create fake images and media content known as deepfakes.

The generator and discriminator refer to the two neural networks that make up a GAN. A convolutional neural network serves as the generator, while a deconvolutional neural network serves as the discriminator. The generator's objective is to generate outputs that should easily be misinterpreted as real data. The discriminator's purpose is to figure out which of the outputs it gets were generated intentionally. Figure 2.1 shows the basic architecture of GANs.

When training begins, the generator generates bogus data, which the discriminator soon recognises as being such. The discriminator's classification gives a signal that the generator uses to change its weights via backpropagation.

As shown in Figure 2.2, the discriminator's training data is derived from two sources: genuine data, such as actual paintings (real), and artificial data (fake). During training, the discriminator considers these examples as positive examples. The generator generates fake data. During training, the discriminator uses these situations as negative examples.

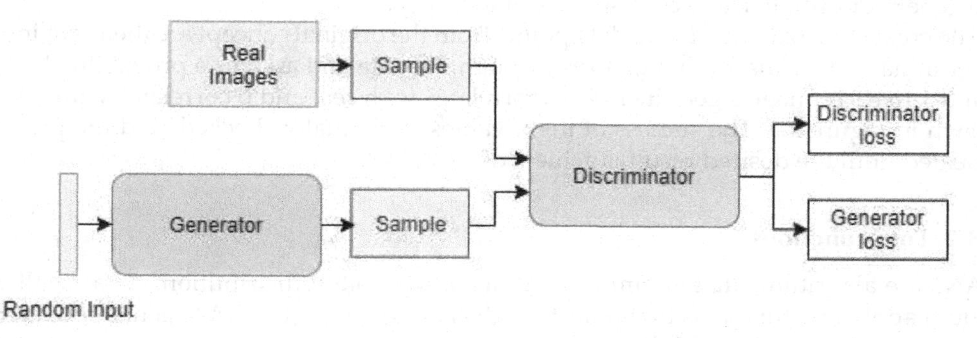


**FIGURE 2.1**
GAN architecture.

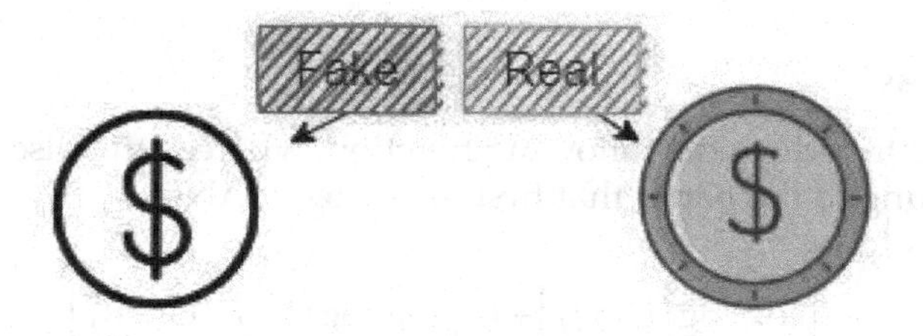


**FIGURE 2.2**
Discriminator classifies between real and fake results.

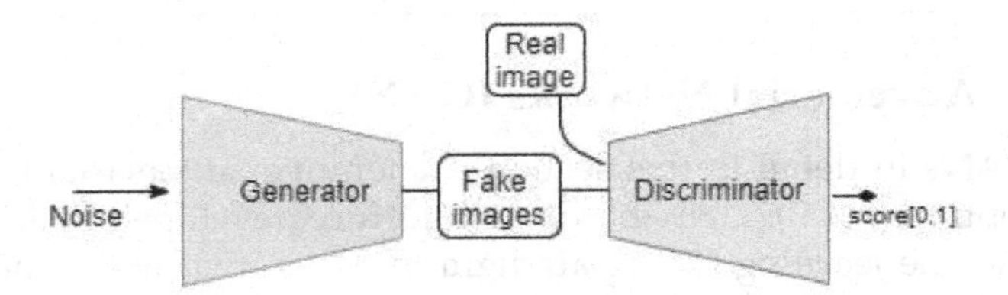


**FIGURE 2.3**
GAN flow diagram.

Let's say we want to create handwritten numerals similar to those in the Modified National Institute of Standards and Technology database (MNIST) [51] collection based on real-world data. When provided an instance from the true MNIST dataset, the discriminator's purpose is to identify genuine ones.

Meanwhile, the generator generates new synthetic images for the discriminator to process. It does so hoping that they, too, will be recognised as genuine, although they are not. The generator's objective is to generate presentable handwritten digits, allowing the user to deceive without being discovered. The discriminator's job is to recognise bogus images generated by the generator.

### 2.3.1 How GANs Work

The first step in building a GAN is to identify the desired end output and generate an initial training dataset based on those parameters. This data is then randomised and fed into the generator until it achieves basic output accuracy.

The created visuals and actual data points from the original concept are then sent into the discriminator. The discriminator sorts through the data and assigns a probability between 0 and 1 to each image's genuineness (1 correlates with real and 0 correlates with fake), as shown in Figure 2.3. The success of these values is manually checked, and the process is repeated until the desired result is achieved.

### 2.3.2 Loss Functions

GANs are algorithms that attempt to imitate a probability distribution. As a result, they should apply loss functions that reflect the distance between the GAN-generated data distribution and the real-world distribution.

One loss function can be used for generator training and the other for discriminator training in a GAN.

#### *2.3.2.1 Minimax Loss*

The generator seeks to reduce the following function, whereas the discriminator strives to maximise it, according to the paper that first presented GANs:

$$E_x\left[\log\log\left(D(x)\right)\right]+E_z\left[\log\log\left(1-D\left(G(z)\right)\right)\right] \tag{1.2}$$

In this function:

The discriminator's estimate of the likelihood that actual data instance $x$ is real is $D(x)$.
The intended value for all genuine data instances is $E_x$.
When given noise $z$, the generator's output is $G(z)$.
The discriminator's estimate of the likelihood that a phoney instance is real is $D(G(z))$.
$E_z$ is the expected value of all the generator's random inputs (in effect, the expected value of all generated fake instances $G(z)$.
The cross-entropy between the real and produced distributions is the source of the formula.
Because the generator cannot directly alter the function's $\log(D(x))$term, minimising the loss is equivalent to minimising $\log(1-D(G(z)))$.

### 2.3.3 Other Generative Models

GANs aren't the only deep learning-based generative models. Under the name GPT, the Microsoft-backed think tank OpenAI has developed a series of sophisticated natural language generation models (Generative Pre-trained Transformer). They launched GPT-3 in 2020 and made it available through an API. GPT-3 is a potent generative language model that can generate fresh human speech in response to suggestions.

## 2.4 Supervised I2I Translation

In this section, we will look at supervised I2I translation, its applications, and some examples. In supervised I2I translation, we translate source images into target domains where we perform training using numerous paired sets of images of the target domain and source domain.

### 2.4.1 Pix2Pix

I2I translation is essentially the conversion of images from one domain to another domain. Pix2pix [1] is the earliest image translation model. The simple application demonstrating the ability of the pix2pix was an edge map-to-photo creation.

The pix2pix model belongs to the classes of models called paired I2I translation models. The paired I2I translation models use source images and target images; hence they are named. Several image processing and computer vision problems can be solved with the paired I2I translation models. These applications include photoediting tasks to enhance the usability of tasks like background removal and semantic segmentation, which is useful in many applications like image colourisation and autonomous driving tasks. The pix2pix model aims primarily to solve the two-domain image translation problem and is a strong baseline model for the same. It even outperforms many state-of-the-art models, which make use of unstructured or hardcoded loss functions. The two-domain I2I problem can be solved if the source–target images contain the right type and amount of data.

The common framework is proposed by the pix2pix model for two-domain I2I tasks. In the architecture of the pix2pix model, the model uses standard Convolution-BatchNormalisation-ReLU based blocks of layers to create a deep CNN. The U-Net architecture is used for the generator model rather than the traditional encoder–decoder model, which involves taking the source image as an input and down-sampling it for a few layers until a layer wherein the image is up-sampled for a few layers, and a final image is outputted. The UNet architecture also down-samples the image and up-samples it again but would have skip-connections between the encoder and decoder layers of the same size, which would allow the information to be shared between the input and output. The discriminator model takes an input image and a translated image and predicts the likelihood of the translated image as real or a generated image of a source input image. Pix2Pix GAN uses a PatchGAN rather than a DCNN. This aids it in classifying image patches as real or fake instead of an entire image. The discriminator is convolutionally run across the image wherein we average all the responses to give the final output. The network outputs a single feature map of real and fake predictions averaged to give a single score. A 70 × 70 patch size is considered to be effective across different I2I translation tasks. The adversarial loss and the L1 loss between the generated translated image and the expected image are used to train the generator model.

#### 2.4.1.1 Applications of Pix2Pix Models

1. *Texture synthesis and transfer of the texture*: A scenario in which a model takes a texture sample and outputs an infinite amount of picture data that is not identical to the original image but appears to have a similar texture to the eye. It also has the ability to transfer texture from one item to another (e.g., the ability to cut and paste visual texture properties on arbitrary objects). Pix2pix is very good at this problem and generates convincing textures.
2. *Image filters*: Two images can be provided as training data, and the network tries to capture the analogy between these two images. The learned representation is then used as a filter to some new target image to generate the analogous image to that of the given image. This effect can be used for blurring or embossing and also improving the texture of the target image.
3. *Removing deformities from the photographs and photograph enhancements*: It is very useful in any photo refinement and creative application to have the ability to remove the background of the photograph given the person's portrait image and the noise in the photo or colourisation of the black and white photograph or filling in the missing pixels in the photographs. This can be formulated as two-domain image to the image translation problem. Since pix2pix provides us with the general-purpose framework for paired I2I tasks, it does perform very well on the photograph enhancement tasks.
4. *Data denoising*: The purpose of data denoising is to recover the original data from the noisy environment. The simple image translation network can remove the coherent as well as incoherent noise from the data. The training data fed into the pix2pix model is identical clean data-noisy data image pairs. It is compelling only so far as users have access to identical pairs.
5. *Artistic image generation*: The use of GANs and particularly of I2I methods can produce very surreal and dreamy-looking pictures.
6. *Sketch to photo*: A system that could generate a realistic picture from simple freehand sketch drawing without further user input. The system automatically recognises the user intent and produces the artistic-looking image sampled from the training distribution.
7. *Hallucinatory image generation*: Let's say if we provide the image of the landscape during the daytime, and then the network tries to produce the image for the same landscape during the nighttime or during the fog, or during the evening time.

8. *Style transfer*: Style transfer refers to separating and then again combining the content and style of the given natural image. This allows us to create images of high visual quality that can combine the appearance of artwork or scenery. Since the pix2pix model can learn the underlying linear representation in feature space, the separation and combination of object attributes are achieved easily.
9. *Segmentation* [50]: It is one of the foundational tasks in computer vision. It can be defined as the process of dividing the image into regions of related content. The pix2pix model creates the highly complex image vector space and then identifies underlying structures; hence, it can also be used for segmentation tasks.
10. *Scene segmentation*: Scene segmentation essentially understands the objects and physical geometry in a particular scenario. It is well known that there are many segmentation networks, which can segment objects. Still, the most important thing that they do not consider is to have segmentation networks such that their segmented objects look like ground truths from the perspectives of shapes, edges, and contours. The most crucial information of an object is the object boundary and the shape and edges of a segmented building. It can be controlled or improved with the help of contours. So after adding corresponding contours to objects of the ground truths, source images and ground truths with overlaid contours form the dataset used in the training of the Pix2Pix network. The FCN-8s architecture can be used for segmentation purposes.

    For problems where the result is less complex than the input, such as semantic segmentation, the GAN seems to have less per-pixel and per-class accuracy than the simple L1 loss and regression. The paper's authors seem to achieve modest accuracy on the segmentation task; It is not very impressive compared with the state-of-the-art. Hence it can be concluded that the pix2pix model is not very suitable for the segmentation tasks.
11. *Semantic labels to photo*: Given semantic labels to generate the high-resolution photos.
12. *Map to aerial photo*: Given map snapshots to produce the aerial view photos.
13. *Architectural labels to photo*: given architectural labels to generate the building photos.
14. *Thermal to colour photos*: Given the thermal view of the image to produce the colour photos.

All of the above problems can be formulated as the paired image-to-image translation tasks. Given the complex inputs, the pix2pix model produces the desired outputs conditioned on some data.

### 2.4.2 Additional Work on Supervised I2I Translations

#### *2.4.2.1 Single-Modal Outputs*

With one-to-one correspondence, single-modal I2I converts the input data from one domain to an output in the target domain. In this section, we will discuss different techniques used to achieve this objective.

1. *Image Analogies:*

   The prior work of I2I translation was done by Hertzmann et al. [47] in 2001. One image and its filtered version were presented as training data to create an "analogous" filtered result.

2. *I2I translation with conditional adversarial networks—Pix2Pix:*
By presenting pix2pix in 2016, Isola et al. [1] became the first to use conditional GAN for general-purpose I2I. Between the underlying truth and the translated image, Pix2pix implements pixel-wise regression loss L1. Loss of adversarial training makes sure that the outputs are indistinguishable from "real" photos. This enables it to learn and train the mapping from the input image to the output image.
3. *Pix2PixHD:*
According to Wang et al. [45], results of conditional GANs are limited to low resolution and are far from realistic. In 2017, they introduced an HD version of Pix2Pix, generating results with a resolution of 2048 × 1024 with additional adversarial loss, multi-scale generator, and discriminator architectures. They also included two new functionalities for interactive visual manipulation: 1. Object instance segmentation. 2. Generating a variety of results with the same input. This helped the model achieve outstanding results, where the synthesised images are generated just by taking the sketch (edges image) as an input.
4. *SelectionGAN:*
Tang et al. [43] were the first to attempt multi-channel attention selectionGAN in 2019. It entails the translation of images from differing viewpoints with extreme distortion. SelectionGAN consists of two stages: 1. Target semantic map and the condition image are fed; 2. A multi-channel attention selection mechanism is used to refine the first stage results.
5. *Spatially Adaptive Normalisation (SPADE):*
Park et al. [42] argue that the method of feeding semantic layout directly as an input to the deep network tends to "wash away" semantic information. In 2019, they proposed using input layout for modulation of activations in normalisation layers. The user can get real-time translated images using this model.
6. *Related research work:*
In 2017, by claiming that the loss functions used in pix2pix (pixel-wise regression loss and GAN loss) lead to unclear images, Wang et al. decomposed this task in three steps. 1. Generate images with global structure and local artefacts. 2. Detect the fakest regions from the generated image. 3. Using a revisor, implement "image inpainting" [46]. In 2019, Zhu et al. [41] addressed the shortcomings of SPADE. – It controls the complete picture style with just one style code and only inserts style information at the beginning of the network. Each semantic region's style can be controlled separately using SEAN normalisation. It allows changing segmentation masks or the style for any specified region to alter photographs interactively. According to Shaham et al. [40], traditional I2I designs (Pix2Pix, Pix2PixHD, SPADE, and others) suffer from high processing costs when working with high-resolution images. They proposed an extremely lightweight generator for high-resolution I2I translation in 2020. This model employs pixel-wise networks, in which each pixel is processed independently, resulting in a model that is 18 times faster than the current benchmark. An exemplar-based I2I framework was introduced by Zhang et al. [39] in April 2020, where a photorealistic image could be generated from an input image, given an exemplar image. It jointly learns the cross-domain correspondence and the I2I translation. When calculating a high-resolution relationship, CoCosNet's semantic matching procedure may result in a prohibitive memory footprint. Zhou et al. [38] developed a GRU-assisted refinement module in December 2020, and they were the first to learn the max-resolution, 1024 × 1024, cross-domain semantic correspondence. AlBahar et al. [44] published a paper about guided I2I translation, in which the user controls

the translation while taking into account limitations imposed by a guiding image provided by an external third-party user.

#### *2.4.2.2 Multimodal Outputs*

Multimodal I2I converts data to a target domain distribution of plausible outputs while remaining true to the input. These outputs convey a variety of colour or texture themes (i.e., multimodal) while retaining the original data's semantic meaning. As a result, we consider multimodal I2I to be a unique two-domain I2I. In this section, we will discuss different models which aim to achieve this objective.

1. *BicycleGAN:*
   With the combination of cVAE-GAN and cLR-GAN, BicycleGAN was the first supervised multimodal I2I model to systematically study the solutions to prevent the many-to-one mapping from latent code to the final output during training, also known as the mode collapse problem, and generate realistic outputs [37].
2. *PixelNN—Example-based Image Synthesis:*
   Bansal et al. [36] presented a nearest-neighbour (NN) approach to generate photo-realistic images from an "incomplete," conditioned input to multiple outputs. They addressed two problems in their work. (1) inability to generate a large set diverse output due to mode collapse problem; (2) inability to control the synthesised output. PixelNN was designed in two stages: (1) A CNN to map the input; (2) A pixel-wise nearest neighbour method to map the smoothed output in a controllable manner.
3. *I2I Translation for cross-domain disentanglement:*
   Gonzalez-Garcia et al. [35] divided two domains' internal representation into three sections: the shared portion, which contains information for both domains, and the other two exclusive parts, which represent the domain-specific determinants of variation. They use a new network component called bidirectional image translation based on GAN and a cross-domain autoencoder, which allows bidirectional multimodal translation and retrieval of similar pictures between domains and domain-specific transfer and interpolation between two domains.

## 2.5 Unsupervised I2I (UI2I) Translation

Paired training data isn't accessible for most tasks because:

- Obtaining paired training data can be time-consuming and costly.
- Acquiring input–output pairings for graphic activities such as creative stylisation might be more challenging because the desired output is typically complicated, requiring artistic authoring and supervision.
- The desired result is not well-defined for many tasks, such as object transfiguration (e.g., producing an image of a horse from a supplied image of a zebra).

### 2.5.1 Deep Convolutional GAN (DCGAN)

The primary idea of the DCGAN [5] compared to the general multi-layer perceptron-based generative model is to add the upsampling convolutional layer between the random

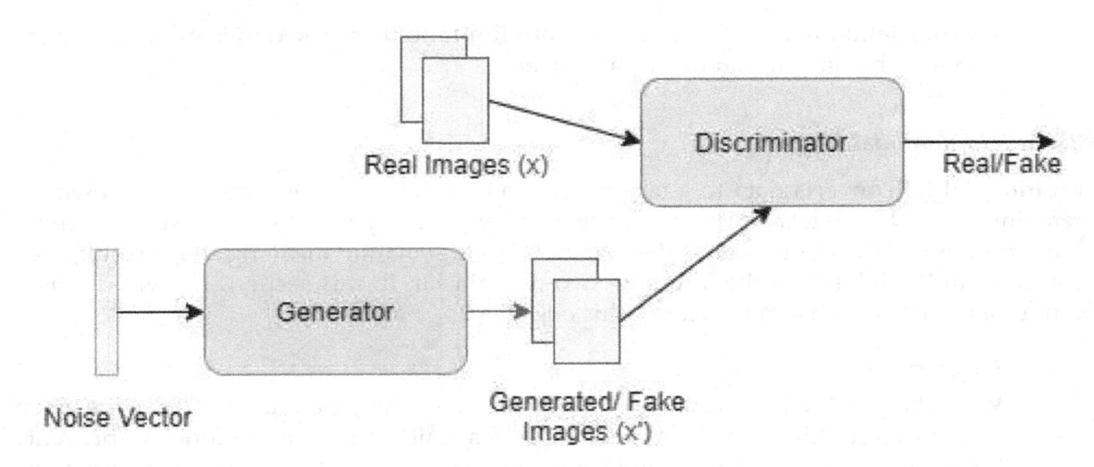


**FIGURE 2.4**
The general architecture of DCGAN.

noise Z and the output images. Figure 2.4 shows the basic structure of DCGAN. In addition, it uses convolutional neural networks to classify the generated and real images as the corresponding label. The authors' inspiration for the DCGAN is to have all the convolutional layers and eliminate fully connected layers. This approach uses $2 \times 2$ kernels or filters, which are foundational to popular convolutional neural networks like AlexNet and LeNet. These kernels have functions like max-pooling and average pooling to group similar pixels together and decrease the spatial resolution by a factor of 2 for $2 \times 2$ kernel size. A $32 \times 32$ image would be processed into a $16 \times 16$ sized image.

Instead of deterministic pooling functions like max-pooling or average pooling, DCGAN uses strided convolutions and helps the network learn its own spatial downsampling. The batch normalisation (reference here) takes a vector feature or a matrix and standardises the inputs to a layer for each mini-batch. It helps the gradient to flow deeper in models and has a stabilising effect on the learning process. The batch normalisation is used in both the generator and discriminator. The activation function used in the generator of the DCGAN is the ReLU activation function except for the output, which uses Tanh and the discriminator uses LeakyReLU activation for all layers.

The DCGAN training methodologies are successful in producing high-resolution images within a stable learning environment. Also, the learned feature representations are reusable and can be used for feature extraction and other downstream tasks. It means that the underlying linear structure in representation space can be leveraged, and object attributes can be generalised for the various I2I translation tasks.

#### *2.5.1.1 DCGAN Applications*

1. Anime faces generation using DCGAN: Animated characters have been a part of everyone's childhood. Most people enjoy watching these characters but only a few can draw them. It requires multiple iterations of sketches to finalise the face of a character. Realistic anime faces can be generated using basic DCGANs. It is one of the applications of deepfakes using GAN. For this process, a DCGAN is trained

using a set of anime face images. The DCGAN models have shown impressive results, which can be used for production purposes.

2. Text-to-image generation using GAN [2]: What if you can generate images just by giving a brief description about it? Well, GANs can do it. GANs have made remarkable progress in generating images of a specific category using a textual description. In recent work, the expressiveness of GANs has enabled capturing and generalising the semantics of input texts, which has helped in improving the performance of the text-to-image generation significantly.
3. Human face editing: Starting from enhancing faces to generating deepfakes, GANs have made a lot of progress in human face editing and generation applications. GANs can generate human faces that do not exist. There are websites based on this concept. Face filters are also developed using GANs, making you look old or completely change your appearance as per the requirements. There are a lot of applications based on this concept. These models can show you how you will look when you get old.

Other applications of DCGAN include 3-D object generation, object extraction, object removal, super-resolution images, real-time texture synthesis, and style transfer.

### 2.5.2 Conditional GAN (CGAN)

Just as we would translate languages using deep neural networks, which are essentially mapping from space to the output space, it is possible to obtain a similar mapping from the source input to the output image, for example, edge map-to-RGB image translation. Conditional GAN [1] aims to solve this problem through the use of conditional modelling. So, the conditional model learns the loss that attempts to determine whether the output image is real or not conditioned on an input image while simultaneously training the generative model to minimise this loss. It simply predicts pixels from pixels conditioned on the given input image. The general architecture of CGAN is shown in Figure 2.5. As we can see, additional conditioned input $Y$ is provided to the generator.

In this case, the generator of the conditional GAN model is generally U-Net-based architecture. The architecture is symmetric and comprises two essential parts: the contracting path, which is made up of the basic convolutional process, and the expansive path, which is made up of transposed 2D convolutional layers. Hence, it can localise and distinguish

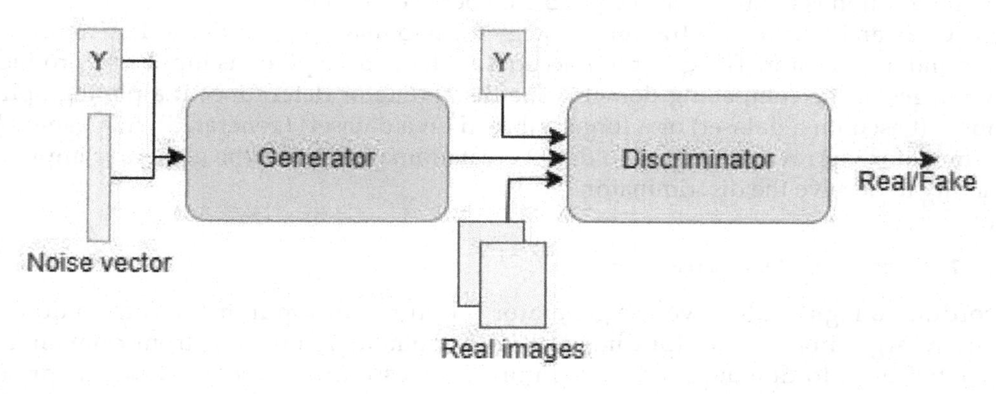


**FIGURE 2.5**
The general architecture of CGAN.

borders by classifying every pixel so the input and output share the same size. For discriminator, the patchGAN based classifier is used. PatchGAN classifier maps input image to $N*N$ array of outputs and classifies whether patch in the image is real or fake, and finally, the results of each patch are averaged. The generator and discriminator both use modules of the form convolution-BatchNorm-ReLu. In addition, the skip connections are added to the general structure of the U-Net, and each skip connection concatenates all the previous layers to the current layer.

Conditional GAN makes use of structured loss to penalise the structure of the output. The important feature of this loss is that it is a learning loss, and hence it does not forcibly condition the output on the input.

The objective of a conditional GAN is expressed as:

$$L_{cGAN}(G,D) = E_{x,y}\left[\log\log\left(D(x,y)\right)\right] + E_{x,z}\left[\log\log\left(1 - D(x,G(x,z))\right)\right] \tag{2.3}$$

where $G$ is attempting to limit the target, whereas $D$ is attempting to maximise it.

The more traditional loss is combined for the image to be close to the ground truth. The CGAN paper uses L1 distance for less blurring.

### 2.5.3 Cycle GAN

Traditionally, a dataset of paired examples is required to train an I2I translation model. That is an extensive dataset including numerous instances of input images $X$ (e.g., summer landscapes) and then the same image modified as an expected output image $Y$ (e.g., winter landscapes).

A paired training dataset is required for this. Preparing these datasets, such as images of different scenes under diverse environments, is difficult and expensive. In many cases, such as iconic artworks and their accompanying pictures, the datasets simply do not exist. CycleGAN [4] is a successful method for unpaired I2I translation. It is an approach to training I2I translation models using the GAN model architecture.

GANs are aimed at creating a zero-sum game between two users: a discriminator and a generator. A differentiable function is used to represent each player, which is managed by a number of parameters. The trained discriminator $D$ discerns between actual and bogus photos, while generator $G$ attempts to make bogus but believable images. In this competition, the solution is to find a Nash equilibrium between $D$ and $G$.

The GAN architecture is a training strategy for two image synthesis models: a discriminator and a generator. The generator accepts a latent space point as input and produces new images in the compelling domain. The discriminator determines if a photograph is genuine (based on a dataset) or a forgery (based on a dataset) (generated). In a game, the discriminator improves its ability to detect created images, while the generator improves its ability to deceive the discriminator.

#### *2.5.3.1 Cycle Consistency Loss*

According to Figure 2.6, if we use generator $G$ to turn our input image $A$ from domain $X$ into a target image or output image $B$ from domain $Y$, image $B$ from domain $Y$ is translated back to domain $X$ by Generator $F$, the essential component of this project is cycle consistency loss. The difference between these two images is referred to as

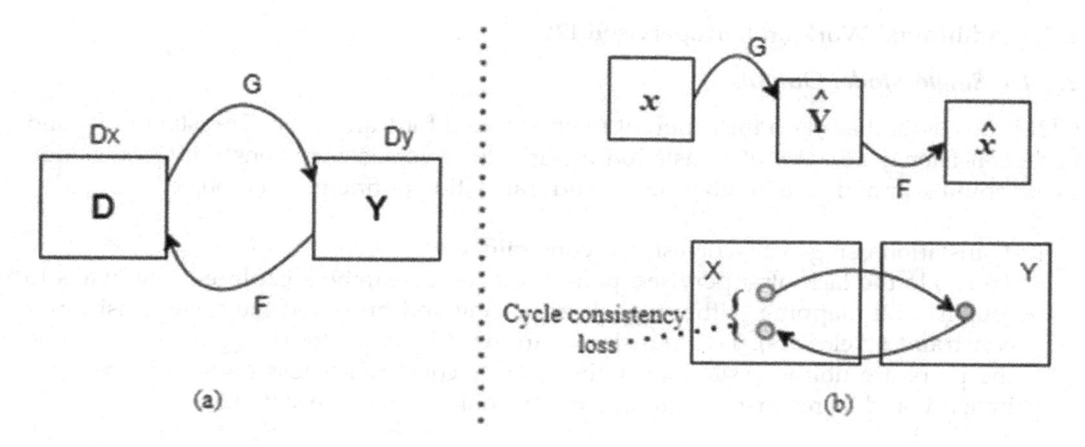


**FIGURE 2.6**
Illustration of cycle consistency loss.

**Source:** [4]

the cycle consistency loss at that time. This approach requires creating two pairs of generators and discriminators: A2B (source to output conversion) and B2A (output to source conversion).

### *2.5.3.2 CycleGAN Applications*

1. Colouring black and white photos: Old photos captured when colour cameras did not exist can be converted to coloured photos using a cycleGAN. Adding colour to a monochromatic photo is a complicated process that requires spending hours paying attention that every pixel in the photo is exactly in its right place. Still, since the introduction of I2I translation models, this process has become much easier.
2. Collection style transfer: Traditional style transfer methods just transfer a particular image style to another. CycleGAN can learn to mimic the style of a complete collection of image artworks instead of just transferring the style of one selected piece of image art. It includes object transformation, object transfiguration, photogeneration from paintings, and season transfer.
3. Photo enhancement: CycleGAN can enhance photo quality, increase resolution (super-resolution), change the depth of field of images, converting regular portraits to DSLR-like images.
4. Virtual try-on [48,49]: Deciding what clothes to buy and how they fit us well takes time. The same applies to sunglasses and jewellery. Trying out outfits, sunglasses, and jewellery is possible using CycleGAN models. It is one of the major applications of CycleGAN, which is currently under research.

Other applications include converting photos from the night to day, translation of satellite images to Google Maps, denoising images, text-to-image translation, and semantic image segmentation [50].

### 2.5.4 Additional Work on Unsupervised I2I

#### 2.5.4.1 Single-Modal Outputs

UI2I is discussed based on the different methods used for translation: Translation beyond cycle consistency constraint, translation using a cycle-consistency constraint, translation by combining knowledge in other fields, and translation of fine-grained objects.

1. Translation using cycle-consistency constraints:
   To tackle the lack of supervised paired images, researchers explored new ways to establish I2I mapping without labels or pairing and proposed the cycle-consistency constraint (cycle loss). Two translators are used in cyclic loss,$G_{A\to B}$, and$G_{B\to A}$, when the pairs are not accessible, to define a cycle-consistency loss among the original image $x_A$and its reconstruction $x_{ABA}$ and the objective can be stated as:

$$L_{\text{cyc}} = L\left(x_A, G_{B\to A}\left(G_{A\to B}\left(x_A\right)\right)\right) \tag{2.4}$$

   *A. Domain Transfer Network (DTN)*
   Taigman et al. introduced a domain transfer network (DTN) for unpaired cross-domain picture synthesis in 2016, which could synthesise images of the target domains' style while keeping their characteristics by assuming constant latent space among two domains [34].

   *B. DualGAN*
   The DualGAN technique, which is based on dual learning from natural language translation, allows image translators to be trained from two sets of unlabelled photos from two domains. In 2017, Zili Yi et al. [33] presented DualGAN, which simultaneously trains two cross-domain transfer GANs with two cyclic losses.

   *C. DiscoGAN*
   In 2017, Taeksoo Kim et al. [32] tackled the problem of detecting cross-domain relationships using unpaired data. DiscoGAN learns to recognise relationships between domains which it then employs to transfer style from one domain to another while maintaining key characteristics.

   *D. CycleGAN*
   Isola et al. [4] presented a method for learning to translate a graphic image from the source to the target domain without paired examples in 2017.

   *E. Unsupervised I2I Translation (UNIT)*
   UNIT was built by Liu et al. [31] to test the shared latent space assumption, which asserts that in a shared latent space, the same latent code can be mapped to a group of corresponding pictures from different domains. They demonstrated that the shared-latent space constraint implies the cycle-consistency constraint.

   *F. Related research work:* Li et al. [30] contend that I2I translation using GAN with cycle-consistency loss may not produce plausible results when the input images are high resolution or if the visual gap between two distinct domains is substantial. They proposed SCANs by deconstructing a single translation into multistage transformations, which strengthens high-resolution I2I translation. Kim et al. [29] proposed two novel features in 2019: (1) New attention module; (2) New AdaLIN (Adaptive Layer-instance Normalisation) functions in U-GAT-IT architecture.

They claimed that previous attention methods failed to control geometric changes between domains.

2. Translation beyond cycle-consistency constraint:
   It has successfully eliminated the dependence of paired images for supervised I2I translation. However, it restricts the translation to contain the data of the source input image for reconstruction purposes. It is leading to the inability of homogenous and large domain gap translation limitations. To overcome these limitations, translation beyond the cycle-consistency constraint supports UI2I translation like a cat to the human face, pencil to fork, etc.

   *A. Improving shape deformation in unsupervised I2I translation*
   In 2018, with the idea of semantic segmentation, Gokaslan et al. [28] proposed GANimorph with dilated convolution in the discriminator, which enables training a more context-aware generator.

   *B. TraVeLGAN*
   In 2019, Amodio et al. [27] addressed the limitation of homogenous domain I2I translation. They proposed a three-network system by adding a Siamese network to traditional GAN that directs the generator so that each original image and its generated version have the same semantics. It eliminates the dependency of cycle-consistency constraint, and hence more sophisticated domain mappings can be learned by the generators.

   *C. TransGaGa*
   In 2019, Wu et al. [26] focused on I2I translation across large geometric variations by presenting a new disentangle-and-translate framework. The output images are completely translated and do not have any original characteristics of the input image.

   *D. DistanceGAN*
   One-sided unsupervised domain mapping: Benaim et al. [24] achieved one-sided mapping that keeps the distance between two samples constant. It leads to preferable results over cycle-consistency constraints.

   *E. Geometry-Consistent GAN (GcGAN)*
   Fu et al. [23] argue that the cycle consistency and distance preservation ignore the semantic structure of images. They preserve the given geometric transformation by using geometric consistency constraints.

   *F. Contrastive Learning for Unpaired I2I Translation (CUT)*
   To reflect the content of the input's patch to its corresponding output patch, in 2020, Park et al. [20] used a contrastive learning paradigm to maximise the mutual information between the two.

   *G. High-Resolution Photorealistic Image Translation in Real-Time*
   A Laplacian pyramid translation network: Recently, in 2021, Lian et al. [21] focused on speeding up high-resolution photorealistic I2I translation. They proposed an LPTN (Laplacian pyramid translation network) to translate 4K resolution images in real-time with a GPU.

   *H. Swapping Autoencoder for Deep Image Manipulation*
   Park et al. [18] proposed a swapping autoencoder model that allows controllable manipulation of images rather than random sampling.

I. *Related research work*
Katzir et al. [25] claim that the shape translation task is far more challenging than the transfer of style and appearance due to its complex, non-local nature. They overcame this challenge by descending the deep layers of VGG-19 and applying the translation from and between these deep features. Jiang et al. [19] proposed a simple two-steam generative model with new feature transformation in a coarse-to-fine fashion. The settings are very clean and simple without any cycle-consistency constraints. Recently, to overcome the drawbacks of cycle-consistency loss, Zhao et al. [12] proposed ACL-GAN, which uses a new adversarial-consistency loss as a replacement to the cycle-consistency loss to retain important features of the source image in 2020. In 2021, Zheng et al. [22] proposed a new spatially correlative loss that preserves scene structure consistency with large appearance changes. They believe that pixel-level cycle consistency hinders translation across large domain gaps.

3. Translation of fine-grained objects:
Translations that use or beyond cycle consistency focus on global style transfer, for example, summer-to-winter translation. Whereas in some applications, we have to focus on local objects, for example, virtual try-on, in this case, the focus of translation must be shifted to fine-grained objects.

A. *Unsupervised Attention-guided Image-to-Image Translation*
Mejjati et al. [15] presented an unsupervised attention mechanism to focus the attention of I2I translation on individual objects.

B. *InstaGAN*
Mo et al. [14] addressed multiple target instances translation and translation with a drastic change in shape. They proposed a context-preserving loss that learns the identity function outside of the target instances. The results show that the model outperforms the CycleGAN mode.

C. *Related research work*
Chen et al. [16] successfully decomposed the task into two separate networks: (1) Attention network predicts spatial attention map of images. (2) Transformation network focuses on translating objects as discussed in *Attention-GAN for Object Transfiguration in Wild Images*. In 2018, Ma et al. [17] proposed a new framework for the purpose of instance-level I2I by using deep attention (DAGAN), which assists in breaking down the translation into instance level. In 2019, Shen et al. [13] presented a simple instance-aware I2I translation approach (INIT), which focuses on fine-grained local objects as well as global styles. They proposed the instance-level objective loss for a more precise reconstruction of objects.

4. Combining knowledge from different fields for image translation:
By combining knowledge of some other research areas, it is possible to improve the effectiveness of I2I translating models.

A. *Art2Real*
In 2018, Tomei et al. [11] developed a semantic-aware architecture that translates artistic artworks to realistic images.

B. *Related research work*
Cho et al. [10] proposed group-wise deep whitening-and-colouring transformation (GDWCT), extending whitening-and-colouring transformation (WCT) with their new

regularisation methods. This model is quite good at capturing the exemplar's style. Recently, in 2020, the knowledge distillation method, where a heavy teacher generator trains a student generator with low-level information was used by Chen et al. [8]. A student generator is included to measure the length between real images. Chen et al. [6] claim that the existing unsupervised I2I translations lack generalisation capacity, that is, they fail to translate images that are out of their training domain. They proposed two sub-modules: (1) To validate testing samples to input domain; (2) To transform the output to the expected results. Ouderaa et al. [9] interpolate invertible neural networks (INNs), which partially supports cycle-consistency into I2I to increase the fidelity of the output and reduce the memory overhead (RevGAN). Chen et al. [7] presented a new role for discriminator by reusing it to encode the target domain images. It results in a more compact and efficient architecture for translating images.

## 2.6 Semi-Supervised I2I

A small number of human-labelled instances for guidance and a large number of labelled examples for automatic translation are required for semi-supervised I2I translation. By combining unpaired and a few paired data, this method can greatly increase translation performance.

Mustafa et al. [52] developed a semi-supervised learning strategy for the first time in a two-domain I2I context due to a lack of labelled data. They introduced transformation consistency regularisation (TCR) as a way to influence a model's prediction to stay the same for geometric transform (perturbed) input samples and their reconstruction versions. Unlabelled data along with less than 1% of labelled data can be used in their method.

## 2.7 Few-shot I2I

Humans have a quick learning ability, which allows them to learn from very few examples and achieve exceptional accomplishments. With just a few sample photographs, a toddler can learn the difference between "Lion" and "Tiger." Existing I2I models, on the other hand, are unable to translate images from only a few source and destination domain training examples. Researchers assume that if a model has learned a large quantity of data in one area, it will learn a new category rapidly with only a few examples. With only a few samples, a Few-shot I2I model can transfer or generate images.

Transferring GAN (TGAN) was introduced by Wang et al. [53], which combines transfer learning with GAN by applying information to a target domain using a pre-trained network trained on a source domain with only a few images. This method reduces the convergence time and improves the quality of the generated images greatly. To overcome the challenge of one-shot cross-domain translation, Benaim et al. [56] presented one-shot translation (OST), which tries to train a unidirectional mapping function given a series of images from the target domain and a source image. Cohen et al. [55] offer BiOST as an extension of OST, which uses a feature-cycle consistency term to translate in both directions without using a weight sharing technique.

## 2.8 Comparative Analysis

We compare the results of seven I2I translation models on the edge-to-shoes translation task in this section. The official public version is used for all experimental codes. The UT-Zap50K dataset [57] is used to assess the efficiency of two-domain I2I methods. This dataset contains 49K+ image pairs, each of which contains a shoe image and its corresponding edge map. Images from the source and target domains are not paired in an unsupervised setting.

### 2.8.1 Metrics

The results are evaluated based on image quality and image diversity using inception score (IS), Fréchet inception distance (FID), and learned perceptual image patch similarity (LPIPS).

- The inception score (IS) [59] determines how realistic the output of a GAN is. It measures two things at the same time: the images vary (i.e., the types of shoes vary), and each image clearly resembles something (i.e., each image is distinguishable from other categories). A higher score indicates an improved performance.
- The Fréchet inception distance (FID) [58] is calculated by calculating the variance and mean distances between generated and real photos in a deep feature space. It contrasts the distribution of generated images with that of real ones. An improved performance is indicated by a lower score.
- Learned perceptual image patch similarity (LPIPS) [60] assesses the variety of translated photos and has been shown to correlate well with a human perceptual similarity. It is calculated as the average LPIPS distance between two pairs of translation outputs from the same input that were randomly sampled. A higher LPIPS score indicates a more realistic and diverse translation result.

### 2.8.2 Results

As experimented by Pang et al. [61], the supervised methods pix2pix and BicycleGAN outperform the unsupervised methods CycleGAN, U-GAT-IT, and GDWCT in terms of FID, IS, and LPIPS. The newest technique, CUT, receives the highest FID and IS scores in Table 2.1 without any supervision. We prefer supervised methods over unsupervised methods. There could be several reasons for this. To begin, CUT's backbone, StyleGAN, is a robust GAN for image synthesis in comparison to others. Furthermore, the contrastive learning method they employed is an effective content constraint for translation.

On a similar network model, the outcomes of supervised methods are usually better than those of unsupervised approaches when it comes to translation. Unsupervised methods, on the other hand, like CUT, benefits from developing network architecture (StyleGAN) and more efficient training techniques (contrastive learning), which outperform supervised methods in some cases. As shown in Tables 2.1 and 2.2, selecting a new model for training the I2I task might be a good idea because such a model is typically trained using some of

**TABLE 2.1**

Comparison of the Average IS, FID, and LPIPS Scores of Two-domain Unsupervised I2I Methods Trained on the UT-Zap50K Dataset for Edge-to-Shoes Translation (The Highest Scores are Highlighted)

| | Unsupervised I2I | | | | |
|---|---|---|---|---|---|
| **Category** | **Single-modal** | | | | **Multimodal** |
| Method | GDWCT | U-GAI-IT | CycleGAN | CUT | MUNIT |
| Publication | 2019 | 2019 | 2017 | 2020 | 2018 |
| IS (↑) | 2.69±0.39 | 2.86±0.31 | 2.66±0.25 | **3.21±0.77** | 2.33±0.25 |
| FID (↓) | 79.56 | 91.33 | 73.76 | **50.03** | 75.31 |
| LPIPS (↑) | 0.017 | 0.028 | 0.038 | 0.019 | 0.19 |

Source: [57]

**TABLE 2.2**

Comparison of the Average IS, FID, and LPIPS Scores of Two-domain Supervised I2I Methods Trained on the UT-Zap50K Dataset [57] for Edge-to-Shoes Translation (The Highest Scores are Highlighted)

| | Supervised I2I | |
|---|---|---|
| **Category** | **Multimodal** | **Single-modal** |
| Method | BicycleGAN | pix2pix |
| Publication | 2017 | 2017 |
| IS (↑) | 2.96±0.36 | 3.08±0.39 |
| FID (↓) | 64.23 | 65.09 |
| LPIPS (↑) | **0.237** | 0.064 |

Source: [57]

the most recent training techniques and well-designed network architecture. Furthermore, a high-quality dataset is critical in the I2I task.

## 2.9 Conclusion

I2I translation problems are prominent in computer vision. Partly because the concept of transferring the qualities of one image to another is intuitive to us, and the applications of such translation systems are endless. I2I assignments can be used to formulate and study a wide range of industrial and creative image generating problems. The older, more precise, and limited I2I systems are less practical, require a lot of manual adjusting, and are less visually appealing. With the use of a general-purpose framework, I2I problems can be developed and solved after the development of GANs. In this chapter, we look at GANs' supervised and unsupervised learning methods, as well as their remarkable capacity to learn local image features. We investigate GANs' capacity to generate high-quality images, as well as their ability to generate samples quickly and the flexibility they afford in terms of the loss function and network architecture. In addition to its advantages, GAN has certain drawbacks, including a lack of stability during training and no direct indicator of convergence. We also look into conditional modelling and generation with GANs and labels. In this chapter, we also look at the specific architectures that are designed to address these issues. The relevance of I2I methods and advances in GANs has introduced many interesting applications.

## References


1. P. Isola, J.- Y. Zhu, T. Zhou, and A. A. Efros, "Image-to-Image Translation with Conditional Adversarial Networks," *2017 IEEE Conf. Comput. Vis. Pattern Recognit.*, pp. 5967–5976, 2017.
2. H. Zhang et al., "StackGAN: Text to Photo-Realistic Image Synthesis with Stacked Generative Adversarial Networks," *2017 IEEE Int. Conf. Comput. Vis.*, pp. 5908–5916, 2017.
3. I. J. Goodfellow et al., "Generative Adversarial Networks," *ArXiv*, vol. abs/1406.2, 2014.
4. J.-Y. Zhu, T. Park, P. Isola, and A. A. Efros, "Unpaired Image-to-Image Translation Using Cycle-Consistent Adversarial Networks," *2017 IEEE Int. Conf. Comput. Vis.*, pp. 2242–2251, 2017.
5. A. Radford, L. Metz, and S. Chintala, "Unsupervised Representation Learning with Deep Convolutional Generative Adversarial Networks," *CoRR*, vol. abs/1511.0, 2016.
6. Y.-C. Chen, X. Xu, and J. Jia, "Domain Adaptive Image-to-Image Translation," *2020 IEEE/CVF Conf. Comput. Vis. Pattern Recognit.*, pp. 5273–5282, 2020.
7. R. Chen, W. Huang, B. Huang, F. Sun, and B. Fang, "Reusing Discriminators for Encoding: Towards Unsupervised Image-to-Image Translation," in *2020 IEEE/CVF Conference on Computer Vision and Pattern Recognition (CVPR)*, 2020, pp. 8165–8174, doi: 10.1109/CVPR42600.2020.00819
8. H. Chen et al., "Distilling portable Generative Adversarial Networks for Image Translation," *The Thirty-Fourth AAAI Conference on Artificial Intelligence (AAAI-20)* 2020, pp. 3585–3592.
9. T. F. A. van der Ouderaa and D. E. Worrall, "Reversible GANs for Memory-Efficient Image-To-Image Translation," *2019 IEEE/CVF Conf. Comput. Vis. Pattern Recognit.*, pp. 4715–4723, 2019.
10. W. Cho, S. Choi, D. K. Park, I. Shin, and J. Choo, "Image-To-Image Translation via Group-Wise Deep Whitening-And-Coloring Transformation," Proceeding of the IEEE Computer Society Conference on Computer Vision and Pattern Recognition, June 2019, 10631–10639.
11. M. Tomei, M. Cornia, L. Baraldi, and R. Cucchiara, "Art2Real: Unfolding the Reality of Artworks via Semantically-Aware Image-To-Image Translation," in *2019 IEEE/CVF Conference on Computer Vision and Pattern Recognition (CVPR)*, 2019, pp. 5842–5852, doi: 10.1109/CVPR.2019.00600
12. Y. Zhao, R. Wu, and H. Dong, "Unpaired Image-to-Image Translation Using Adversarial Consistency Loss," *Computer Vision – ECCV 2020*, pp. 800–815, 2020.
13. Z. Shen, M. Huang, J. Shi, X. Xue, and T. S. Huang, "Towards Instance-Level Image-To-Image Translation," in *2019 IEEE/CVF Conference on Computer Vision and Pattern Recognition (CVPR)*, 2019, pp. 3678–3687, doi: 10.1109/CVPR.2019.00380
14. S. Mo, M. Cho, and J. Shin, "InstaGAN: Instance-aware Image-to-Image Translation," 2019 [Online]. Available: https://openreview.net/forum?id=ryxwJhC9YX
15. Y. Alami Mejjati, C. Richardt, J. Tompkin, D. Cosker, and K. I. Kim, "Unsupervised Attention-guided Image-to-Image Translation," in *Advances in Neural Information Processing Systems*, 2018, vol. 31 [Online]. Available: https://proceedings.neurips.cc/paper/2018/file/4e87337f366f72daa424dae11df0538c-Paper.pdf
16. X. Chen, C. Xu, X. Yang, and D. Tao, "Attention-GAN for Object Transfiguration in Wild Images," *European Conference on Computer Vision*, September 2018, pp. 167–184.

17. S. Ma, J. Fu, C.-W. Chen, and T. Mei, "DA-GAN: Instance-level Image Translation by Deep Attention Generative Adversarial Networks (with Supplementary Materials)," *Computing Research Repository*, 2018, pp. 5657–5666.
18. T. Park et al., "Swapping Autoencoder for Deep Image Manipulation," in *Advances in Neural Information Processing Systems*, 2020, vol. 33, pp. 7198–7211 [Online]. Available: https://proceedings.neurips.cc/paper/2020/file/50905d7b2216bfeccb5b41016357176b-Paper.pdf
19. L. Jiang, C. Zhang, M. Huang, C. Liu, J. Shi, and C. C. Loy, "TSIT: A Simple and Versatile Framework for Image-to-Image Translation," Computer Vision – ECCV 2020. ECCV 2020. Lecture Notes in Computer Science, vol 12348. Springer, 2020. https://doi.org/10.1007/978-3-030-58580-8_13.
20. T. Park, A. A. Efros, R. Zhang, and J.-Y. Zhu, "Contrastive Learning for Unpaired Image-to-Image Translation," *European Conference on Computer Vision*, 2020.
21. J. Liang, H. Zeng, and L. Zhang, "High-Resolution Photorealistic Image Translation in Real-Time: A Laplacian Pyramid Translation Network," in *Proceedings of the IEEE/CVF Conference on Computer Vision and Pattern Recognition (CVPR)*, pp. 9392–9400, June 2021.
22. C. Zheng, T.-J. Cham, and J. Cai, "The Spatially-Correlative Loss for Various Image Translation Tasks," in *Proceedings of the IEEE/CVF Conference on Computer Vision and Pattern Recognition (CVPR)*, Jun. 2021, pp. 16407–16417.
23. H. Fu, M. Gong, C. Wang, K. Batmanghelich, K. Zhang, and D. Tao, "Geometry-Consistent Generative Adversarial Networks for One-Sided Unsupervised Domain Mapping," in *2019 IEEE/CVF Conference on Computer Vision and Pattern Recognition (CVPR)*, 2019, pp. 2422–2431, doi: 10.1109/CVPR.2019.00253.
24. S. Benaim and L. Wolf, "One-Sided Unsupervised Domain Mapping," in *Advances in Neural Information Processing Systems*, 2017, vol. 30 [Online]. Available: https://proceedings.neurips.cc/paper/2017/file/59b90e1005a220e2ebc542eb9d950b1e-Paper.pdf.
25. O. Katzir, D. Lischinski, and D. Cohen-Or, "Cross-Domain Cascaded Deep Feature Translation," *ArXiv*, vol. abs/1906.0, 2019.
26. W. Wu, K. Cao, C. Li, C. Qian, and C. C. Loy, "TransGaGa: Geometry-Aware Unsupervised Image-To-Image Translation," in *2019 IEEE/CVF Conference on Computer Vision and Pattern Recognition (CVPR)*, 2019, pp. 8004–8013, doi: 10.1109/CVPR.2019.00820
27. M. Amodio and S. Krishnaswamy, "TraVeLGAN: Image-To-Image Translation by Transformation Vector Learning," *2019 IEEE/CVF Conf. Comput. Vis. Pattern Recognit.*, pp. 8975–8984, 2019.
28. A. Gokaslan, V. Ramanujan, D. Ritchie, K. I. Kim, and J. Tompkin, "Improving Shape Deformation in Unsupervised Image-to-Image Translation," *ArXiv*, vol. abs/1808.0, 2018.
29. J. Kim, M. Kim, H. Kang, and K. H. Lee, "U-GAT-IT: Unsupervised Generative Attentional Networks with Adaptive Layer-Instance Normalization for Image-to-Image Translation," 2020 [Online]. Available: https://openreview.net/forum?id=BJlZ5ySKPH
30. M. Li, H. Huang, L. Ma, W. Liu, T. Zhang, and Y. Jiang, "Unsupervised Image-to-Image Translation with Stacked Cycle-Consistent Adversarial Networks," *15th European Conference, Proceeding, Part IX*, pp. 186–201, September 2018.
31. M.-Y. Liu, T. Breuel, and J. Kautz, "Unsupervised Image-to-Image Translation Networks," in *Proceedings of the 31st International Conference on Neural Information Processing Systems*, pp. 700–708, 2017.
32. T. Kim, M. Cha, H. Kim, J. K. Lee, and J. Kim, "Learning to Discover Cross-Domain Relations with Generative Adversarial Networks," *Proceeding of the 34th International Conference on Machine Learning*, pp. 1857–1865, 2017.
33. Z. Yi, H. Zhang, P. Tan, and M. Gong, "DualGAN: Unsupervised Dual Learning for Image-to-Image Translation," in *2017 IEEE International Conference on Computer Vision (ICCV)*, 2017, pp. 2868–2876, doi: 10.1109/ICCV.2017.310
34. Y. Taigman, A. Polyak, and L. Wolf, "Unsupervised Cross-Domain Image Generation," arXiv preprint arXiv:1611.02200, 2016.

35. A. Gonzalez-Garcia, J. van de Weijer, and Y. Bengio, "Image-to-Image Translation for Cross-Domain Disentanglement," in *Proceedings of the 32nd International Conference on Neural Information Processing Systems*, pp. 1294–1305, 2018.
36. A. Bansal, Y. Sheikh, and D. Ramanan, "PixelNN: Example-based Image Synthesis," *arXiv*, 2017.
37. J.-Y. Zhu et al., "Toward Multimodal Image-to-Image Translation," in *Advances in Neural Information Processing Systems*, 2017, vol. 30 [Online]. Available: https://proceedings.neurips.cc/paper/2017/file/819f46e52c25763a55cc642422644317-Paper.pdf.
38. X. Zhou et al., *Full-Resolution Correspondence Learning for Image Translation, ArXiv*, vol. abs/2012.02047, 2020.
39. P. Zhang, B. Zhang, D. Chen, L. Yuan, and F. Wen, "Cross-Domain Correspondence Learning for Exemplar-Based Image Translation," in *2020 IEEE/CVF Conference on Computer Vision and Pattern Recognition (CVPR)*, 2020, pp. 5142–5152, doi: 10.1109/CVPR42600.2020.00519
40. T. R. Shaham, M. Gharbi, R. Zhang, E. Shechtman, and T. Michaeli, "Spatially-Adaptive Pixelwise Networks for Fast Image Translation," in *Proceedings of the IEEE/CVF Conference on Computer Vision and Pattern Recognition (CVPR)*, pp. 14882–14891, June 2021.
41. P. Zhu, R. Abdal, Y. Qin, and P. Wonka, "SEAN: Image Synthesis with Semantic Region-Adaptive Normalization," in *2020 IEEE/CVF Conference on Computer Vision and Pattern Recognition (CVPR)*, 2020, pp. 5103–5112, doi: 10.1109/CVPR42600.2020.00515
42. T. Park, M.-Y. Liu, T.-C. Wang, and J.-Y. Zhu, "Semantic Image Synthesis with Spatially-Adaptive Normalisation," in *2019 IEEE/CVF Conference on Computer Vision and Pattern Recognition (CVPR)*, 2019, pp. 2332–2341, doi: 10.1109/CVPR.2019.00244
43. H. Tang, D. Xu, N. Sebe, Y. Wang, J. J. Corso, and Y. Yan, "Multi-Channel Attention Selection GAN With Cascaded Semantic Guidance for Cross-View Image Translation," in *2019 IEEE/CVF Conference on Computer Vision and Pattern Recognition (CVPR)*, 2019, pp. 2412–2421, doi: 10.1109/CVPR.2019.00252
44. B. Albahar and J.-B. Huang, "Guided Image-to-Image Translation with Bi-Directional Feature Transformation," in *2019 IEEE/CVF International Conference on Computer Vision (ICCV)*, 2019, pp. 9015–9024, doi: 10.1109/ICCV.2019.00911
45. T.-C. Wang, M.-Y. Liu, J.-Y. Zhu, A. Tao, J. Kautz, and B. Catanzaro, "High-Resolution Image Synthesis and Semantic Manipulation with Conditional GANs," in *2018 IEEE/CVF Conference on Computer Vision and Pattern Recognition*, 2018, pp. 8798–8807, doi: 10.1109/CVPR.2018.00917
46. C. Wang, H. Zheng, Z. Yu, Z. Zheng, Z. Gu, and B. Zheng, "Discriminative Region Proposal Adversarial Networks for High-Quality Image-to-Image Translation," 15th European Conference, Germany, September 8–14, 2018, Proceedings, Part I, Sep 2018 Pages 796–812, https://doi.org/10.1007/978-3-030-01246-5_47.
47. A. Hertzmann, C. E. Jacobs, N. Oliver, B. Curless, and D. Salesin, "Image Analogies," *Proc. 28th Annu. Conf. Comput. Graph. Interact. Tech.*, 2001.
48. D. Song, T. Li, Z. Mao, and A.-A. Liu, "SP-VITON: Shape-Preserving Image-Based Virtual Try-On Network," *Multimed. Tools Appl.*, vol. 79, no. 45, pp. 33757–33769, 2020, doi: 10.1007/s11042-019-08363-w
49. F. Sun, J. Guo, Z. Su, and C. Gao, "Image-Based Virtual Try-on Network with Structural Coherence," in *2019 IEEE International Conference on Image Processing (ICIP)*, 2019, pp. 519–523, doi: 10.1109/ICIP.2019.8803811
50. E. Shelhamer, J. Long, and T. Darrell, "Fully Convolutional Networks for Semantic Segmentation.," *IEEE Trans. Pattern Anal. Mach. Intell.*, April 2017, vol. 39, no. 4, pp. 640–651, doi: 10.1109/TPAMI.2016.2572683
51. Y. LeCun and C. Cortes, "{MNIST} hand-written digit database," 2010, [Online]. Available: http://yann.lecun.com/exdb/mnist/
52. A. Mustafa and R. K. Mantiuk, "Transformation Consistency Regularization- A Semi-Supervised Paradigm for Image-to-Image Translation," *European Conference on Computer Vision (online)*, 2020.

53. Y. Wang, C. Wu, L. Herranz, J. van de Weijer, A. Gonzalez-Garcia, and B. Raducanu, "Transferring GANs: Generating Images from Limited Data," *European Conference on Computer Vision, Germany*, 2018.
54. O. Ronneberger, P. Fischer, and T. Brox, "U-Net: Convolutional Networks for Biomedical Image Segmentation," *18th International Conference on Medical Image Computing and Computer Assisted Interventions*, Germany, 2015.
55. T. Cohen and L. Wolf, "Bidirectional One-Shot Unsupervised Domain Mapping," *International Conference on Computer Vision, Korea*, 2019.
56. S. Benaim and L. Wolf, "One-Shot Unsupervised Cross Domain Translation," *32nd Conference on Neural Information Processing Systems (NeurIPS 2018)*, Montréal, Canada, 2018.
57. A. Yu and K. Grauman, "Fine-Grained Visual Comparisons with Local Learning," in *2014 IEEE Conference on Computer Vision and Pattern Recognition*, 2014, pp. 192–199, doi: 10.1109/CVPR.2014.32
58. M. Heusel, H. Ramsauer, T. Unterthiner, B. Nessler, and S. Hochreiter, "GANs Trained by a Two Time-Scale Update Rule Converge to a Local Nash Equilibrium," in *Proceedings of the 31st International Conference on Neural Information Processing Systems*, 2017, pp. 6629–6640.
59. T. Salimans, I. Goodfellow, W. Zaremba, V. Cheung, A. Radford, and X. Chen, "Improved Techniques for Training GANs," in *Proceedings of the 30th International Conference on Neural Information Processing Systems*, pp. 2234–2242, 2016.
60. R. Zhang, P. Isola, A. A. Efros, E. Shechtman, and O. Wang, "The Unreasonable Effectiveness of Deep Features as a Perceptual Metric," in *2018 IEEE/CVF Conference on Computer Vision and Pattern Recognition*, 2018, pp. 586–595, doi: 10.1109/CVPR.2018.00068
61. Y. Pang, J. Lin, T. Qin, and Z. Chen, "Image-to-Image Translation: Methods and Applications," *ArXiv*, vol. abs/2101.08629, 2021.

# 3

# *Image Editing Using Generative Adversarial Network*

**Anuja Jadhav, Chaitrali Sorde, Swati Jaiswal, Roshani Raut, and Atul B. Kathole**

**CONTENTS**



## 3.1 Introduction

Various issues in image handling, processer designs, and PC representation might outline as "make an interpretation of" cooperation images into a matching yield images. Like how thought might be conveyed in one or the other English or French, a demonstration

DOI: 10.1201/9781003203964-3

can be portrayed as an RGB picture, a slope field, a control map, or a semantic marker map, in addition to other things. Likewise, to robotized language change, we portray customized picture to-picture transformation as the gig of changing over one possible portrayal of a segment into another, predetermined adequate preparation data. Historically, each of these jobs has been accomplished using distinct, specialized gear, despite always the similar goal: forecast pixels from pixels. This study aims to provide an outline for all of these issues. The field has previously made great strides in this regard, with convolutional neuronal networks (CNNs) emerging as the de facto standard for solving a broad range of picture prediction issues. CNNs learn to lessen a loss role—a goal that quantifies the excellence of outcomes—and, though the learning procedure is automated, much human work is required to construct successful losses. In other words, we must continue to inform CNN of the information we desire to have minimized. However, like King Midas, we must exercise caution with our wishes! The results will be fuzzy if we use a primary method and instruct the CNN to lessen the Euclidean detachment between projected and ground reality pixels. This is because the Euclidean detachment is reduced by all possible yields, resulting in distorting. Developing loss roles that compel the CNN to do the task at hand—for example, generate precise, realistic pictures—is an open challenge that often needs specialist expertise. It would be exceedingly lovely if we could in its place define a high-level aim, such as "make the output indistinguishable from reality," and then train a suitable loss role automatically. Fortunately, the newly suggested generative adversarial networks (GANs) accomplish precisely this goal. GANs acquire a loss function that attempts to categorize the yield picture as genuine or false while also training a reproductive model to reduce this loss. Fuzzy photographs will not be accepted since they seem to be manufactured. Because GANs learn an adaptive loss function from the data, they may be used to a wide variety of tasks that would need quite different loss functions in the past.

## 3.2 Background of GAN

GAN is a phenomenal generative model. In any case, the first GAN model had numerous issues, like poor diversity, vanishing gradient, and training difficulty [3]. Numerous endeavors have been committed to getting better GANs through various enhancement strategies. Hence, beginning around 2014, speculations and articles connected with GAN have come out in an unending stream. Numerous new GANs-based models have been proposed to work on the security and nature of the produced outcomes.

Two neural networks challenge one another in a game. Given a preparation set, this procedure figures out how to produce new information with similar insights as to the preparation set. For instance, a GAN prepared on photos can make new photos that glance cursorily valid to human onlookers, having numerous sensible qualities.

The center thought of a GAN depends on the "backhanded" preparation through the discriminator, which is updated progressively. This essentially implies that the generator isn't prepared to limit the distance to a particular picture but trick the discriminator. This empowers the model to learn in an unsupervised way. Setting two calculations in opposition to one another started with Arthur Samuel, a noticeable scientist in the field of software engineering who is credited with promoting the expression "machine learning." While at IBM, he concocted a checkers game—the Samuel Checkers-playing Program—that was among the first to effectively self-learn, to some degree by assessing the shot at each side's triumph at a given position.

In 2014, Goodfellow and his associates provided the main working execution of a generative model dependent on adversarial networks. Goodfellow has frequently expressed that he was roused by contrastive noise assessment, a method of learning information dissemination by looking at it against a noise distribution, i.e., a numerical capacity addressing distorted or corrupted information. The contrastive noise assessment involves a similar loss function as GANs—at the end of the day, a similar proportion of execution concerning a model's capacity to expect anticipated results.

Once more, GANs are comprised of two sections: generators and discriminators. The generator model produces engineered models (e.g., images) from random noise sampled utilizing a distribution model, which alongside genuine models from an informational preparation collection are taken care of to the discriminator, which endeavors to recognize the two. Both the generator and discriminator work in their respective capacities until the discriminator can't perceive the genuine models from the integrated models with better than the half-precision expected of possibility.

GANs are algorithmic plans that use two neural associations, setting one contrary to the following to make new, fabricated instances of data that can be passed with accurate data. GANs train unsupervised, implying that they construe the examples within datasets without reference to known, named, or explained results. Curiously, the discriminator's work illuminates that regarding the generator—each time the discriminator accurately recognizes an incorporated work, it advises the Generator how to change its result so it very well may be more reasonable later on.

## 3.3 Image-to-Image Translation

CycleGAN was used for image-to-image translation undertakings in the nonattendance of combined models. It takes planning from a source area A to an objective space B by presenting two-cycle consistency misfortunes. Also, DiscoGAN and DualGAN utilize an unaided learning approach for image-to-image translation dependent on unsupervised information yet diverse loss functions. HarmonicGAN used for unsupervised image-to-image interpretation presents spatial smoothing to uphold predictable mappings during the performance.

The picture-to-picture interpretation takes in arranging between pictures from a source region and pictures from an objective region. Its applications incorporate picture colorization, semantic names from photographs [4], superresolution pictures, and spatial variety.

Overall, a picture-to-picture interpretation method requires a recognition of spaces of interest in the information pictures and sort out some way to decipher the recognized districts into the objective region. In an unaided setting with no paired pictures between the two spaces, one ought to zero in on the thought spaces that are obligated to move. The task of finding areas of interest is more huge in a picture-to-picture interpretation. The translation ought to be applied particularly to a particular article as opposed to the whole picture.

CycleGANs empower planning to start with one area A then onto the next space B without finding impeccably coordinated, preparing sets. Assume we have many images from area A and an unpaired arrangement of pictures from space B. We need to have the option to decipher an idea starting with one set and then onto the next. To do this, we characterize a planning G(G: A→B) that makes an honest effort to plan A to B. Be that as it may, with unpaired information, we presently can't take a gander at genuine and counterfeit sets of data. Yet, we realize that we can change our model to create a result that has a place with an objective space.

So when you push an image to domain A, we can prepare a generator to deliver beautiful images to domain B. In any case, the issue with that will be that we can't constrain the result of the generator to relate to its input. This prompts an issue called mode breakdown, in which a model may plan different contributions from area A into a similar result from space B. In such cases, we know that the result should resemble domain B given info to domain B. In any case, to properly plan the contribution to the relating objective space, we present extra planning as reverse planning G′(G′: →A), which attempts to plan B to A. This is called the cycle-consistency imperative.

Assuming we decipher an image from domain A to an embodiment of domain B. Afterward, we interpret back from domain B to a part A, we ought to show up around at a similar image of the pony we began.

A total interpretation cycle should carry you back to a similar image you began with. Assuming the accompanying condition is met on image translation structure space A to B, we say that an image change from area A to area B was correct.

$$G_{\mathrm{AtoB}}(G_{\mathrm{AtoB}}(x)) \approx x$$

CycleGAN ensures that the model correctly plans from space A to area B. The steps required to perform image-to-image translation requires the following steps:

1. Visualization of datasets
2. Defining an appropriate model
3. Make use of a discriminator
4. Use of residual blocks/functions
5. Make use of generator
6. Training process
   a. Calculation of discriminator and generator function
   b. Make use of an optimizer to get good results.

## 3.4 Motivation and Contribution

Even though Gregor et al. [6] and Dosovitskiy et al. [7] can make realistic images, they don't use the generators for learning. A deeper convolutional network design makes more progress while stacking two generators to create more realistic images. InfoGAN learns a way to think about things that are easier to understand. There are many ways to train GANs and show them how. They came up with Wasserstein GAN, an alternative way to introduce a GAN that doesn't have to balance the discriminator and generator.

Putting in and motivating with the development of deep learning [8], many new ways to improve the quality of images have been shown. Pictures made with deep generative adversarial networks have recently caught the attention of a lot of academics because they can better represent the probability distribution than traditional generative models. Several research papers have said that using adversarial networks to make pictures is a good idea. On the other hand, existing GANs don't work well for picture editing applications because the outputs aren't high-resolution or realistic enough to be used in these applications. Several researchers [9,10,11,12,13] have looked at generative adversarial networks and GAN variations, and one of the applications is making pictures. Some [14,15] looked at GAN-based image creation and only briefly looked at picture-to-image translation [15], which looked at image synthesis and some picture-to-image translation approaches, which is the most similar to our study. Using GANs to translate images from one picture to another has never been studied before. As a result, this chapter gives an in-depth look at GAN algorithms and variants for translating ideas from one image to another. GAN variants, topologies, and goals are shown, and a general introduction to generative adversarial networks. It talks about the best ways to do image-to-image translation right now, as well as the theory, applications, and issues that are still unanswered. Images can be translated from one format to another in two different ways: supervised and nonsupervised.

## 3.5 GAN Objective Functions

The gap between the actual and produced sample distribution is measured and reduced using an objective function in GANs and variants. Even though GANs have excelled in various tasks, accurate data has presented several challenges. Gradient vanishing and model collapse are two of the functions available. Changing the way things are done will not solve these issues. The structure of GANs is unique. These issues can be solved by reformulating the objective function. As a result, various objective functions have been provided and classified to improve data quality and sample diversity while avoiding the shortcomings of the original GAN and its progeny. Although the GAN design is simple, beginners may find it challenging to grasp GAN loss functions.

### 3.5.1 GAN Loss Challenges

The GAN engineering is described by the minimax GAN misfortune, despite the fact that the non-soaking misfortune work is generally used. Current GANs usually utilize the least-squares and Wasserstein misfortune capacities as substitute misfortune capacities. At the point when different factors, for example, handling power and model hyperparameters are held steady, the huge scope assessment of GAN misfortune capacities shows slight variety.

### 3.5.2 The Problem of GAN Loss

GAN is shown to be very effective, resulting in photorealistic faces, settings, and other items. Although the GAN architecture is straightforward, understanding GAN loss functions might be challenging for beginners.

### 3.5.3 Loss of Discriminator

The discriminator's goal is to maximize the likelihood of real and fake images being assigned. We educate D to give the correct label to both training examples and G samples with the highest chance.

As defined formally, the discriminator seeks to maximize the average log probability for authentic images and the inverted possibilities for phony photos.

$$\log(1 - D(G(z)) + \log(1 - D(G(z))$$

Changes in accordance with model loads would need to be made utilizing stochastic rising as opposed to stochastic plunge if this somehow happened to be carried out straightforwardly. It's normally done as a standard double grouping issue, with names 0 and 1 relating to manufactured and regular pictures, separately. The reason for the model is to lessen the typical paired cross-entropy, frequently known as log misfortune.

### 3.5.4 GAN Loss Minimax

The terms min and max relate to the generator loss minimization and discriminator loss maximization, respectively minimum and maximum (D, G).

The discriminator aims to maximize the average log probability of real photos and the inverse probability of fake images, as described above.

The loss function may not give enough gradient for G to learn well in practice. D can reject samples with high confidence early in learning when G is weak because they are distinct from the training data.

A discriminator and a generator are the two models that make up the GAN architecture. The discriminator categorizes images as real or fake after being taught directly on genuine and manufactured photos (generated). The discriminator model is used instead of now training the generator. The discriminator has been meticulously prepared to produce the generator's loss function. In a two-player game, both the generator and discriminator models fight for improvement simultaneously.

Model convergence on a training dataset is often measured by minimizing the training dataset's selected loss function. The end of a two-player game is signaled by conjunction in a GAN. Instead, the goal is to balance generator and discriminator loss.

## 3.6 Image-to-Image Translation

Recently, GAN-based approaches for image-to-image translation have been extensively employed and have generated some impressive results. Conditional GAN (cGAN) was used in pix2pix to train a plotting from a participation picture to a yield image; cGAN constructs a dependent reproductive model using balancing images from the source and goal domains. Zhu et al. introduced CycleGAN for non-paired image-to-image conversion challenges. They are adding two-cycle consistency losses studies a plotting from foundation domain X to goal domain Y (and vice versa). DiscoGAN and DualGAN use an unsubstantiated learning strategy for image-to-image conversion, albeit with different loss functions. HarmonicGAN, as developed for irregular image-to-image conversion, incorporates spatial leveling to provide continuous mappings throughout the conversion process. Freshly developed unsubstantiated image-to-image conversion network (UNIT) is based on coupled GANs and the shared-latent planetary supposition, which states that two matching pictures from different domains may be plotted to the exact latent depiction.

Additionally, image-to-image conversion approaches presuppose that the latent interstellar of pictures may be partitioned into a gratified and elegant space, allowing for the development of multimodal productions. This established a framework for multimodal unsupervised image-to-image conversion (MUNIT) that incorporates two latent illustrations for style and gratified. The image's content code is coupled with several style illustrations taken from the goal domain to convert a picture to another part. Likewise, Lee et al. presented miscellaneous image-to-image conversion (DRIT) based on the extricated illustration of irregular data, which decays the latent interstellar into two components: a domain-invariant gratified space captures shared data and a domain-specific characteristic space that generates diverse productions given the similar content. The author presented BranchGAN to transmit an image from one area to another using the same distribution and encoder of the two domains. Instagram performs domain-to-domain picture translation on many instances using object segmentation masks. By adding the context-preserving loss, it retains the backdrop. However, since Instagram needs semantic separation labels (i.e., pixel-level explanation) for training, it is limited when such information is not accessible.

The purpose of image-to-image conversion in supervised learning is to discover the plotting between a participation picture and a yield image by training on a collection of matched duplicate pairings. Conditional GAN (cGAN) was used in pix2pix [1] to study a plotting between pairs of pictures from the source and destination domains. BicycleGAN [6] is a model configuration for cGANs that learns a distribution of potential outputs. This approach accomplished the objective of generating multimodal results from a given input.

Unsubstantiated learning requires total independence between source and target picture sets, with no matched instances between the two domains. Due to the high cost and often impossibility of gathering paired training data for a wide variety of applications, there are reasons for unsupervised techniques, such as CycleGAN [2], DualGAN [3], UNIT [4] and UNIT, [5] so on.

### 3.6.1 Controlled Image-to-Image Conversion

#### *3.6.1.1 cGAN*

Isola et al. [17] research conditional argumentative networks (cGANs) as a general-purpose explanation to image-to-image conversion difficulties. They show that their method is helpful for a variety of tasks, including photosynthesis from label maps, object reconstruction from edge maps, and picture colorization. They make available the pix2pix program that was used to create this chapter.

Conditional GANs are used to learn a plotting between an experiential picture $x$ and an arbitrary noise vector $z$ and $y$ in this chapter, $G: \{x, z\} \rightarrow y$. The generator G is taught to generate yields that are indistinguishable from "genuine" pictures by an adversarially proficient discriminator, D, which is qualified to perform as well as feasible in recognizing the generator's "fakes." Figure 3.1 illustrates this working technique. In contrast to an unrestricted GAN, both the generator and discriminator take the input control map into account.

#### *3.6.1.2 BicycleGAN*

Isola et al. [17] intend to describe the delivery of potential productions in a convolutional neural network scenario. They expressly advocate for the invertibility of the relationship

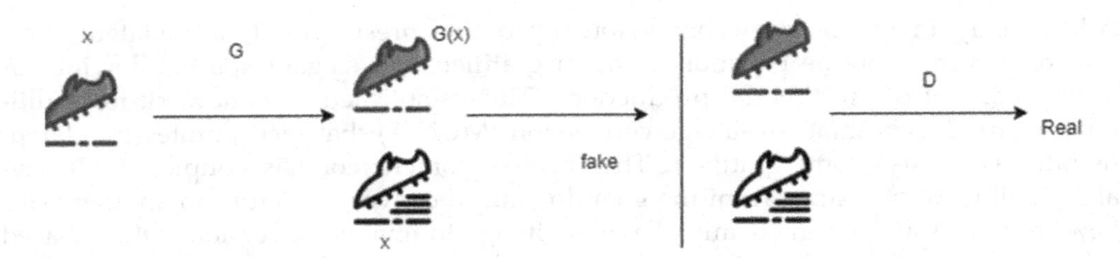


**FIGURE 3.1**
Training a conditional GAN to map edges to photo.

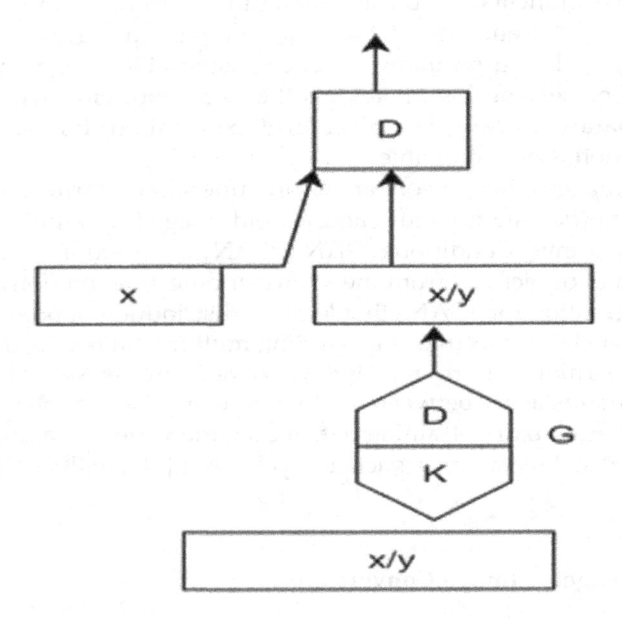


**FIGURE 3.2**
SPA-GAN architecture.

between output and hidden code. BicycleGAN associations cVAE-GAN with aLR-GAN to jointly impose the relationship between dormant encoding and yield, increasing presentation. This supports avoiding a one-to-many plotting between the stationary coding and the product during training, commonly known as the method collapse issue, resulting in more varied outcomes.

#### *3.6.1.3 SPA-GAN*

Emami et al. [4] embedded the attention apparatus directly into the GAN style and presented a unique spatial consideration GAN prototypical (SPA-GAN) for image-to-image conversion. SPA-GAN calculates the discriminator's consideration and uses it to direct the generator's attention to the most discriminatory areas between the foundation and destination areas.

As seen in Figure 3.2, the discriminator creates spatial attention maps and categorizes the input pictures. These maps are provided to the generator and assist it in focusing on the most discriminating object portions. Additionally, the rushed blocks (b-1) and (b-2) illustrate the characteristic map loss, which is defined as the alteration between the characteristic maps of the (joined) actual and produced pictures calculated in the decoder's first sheet. The typical map loss is employed in image translation to maintain domain-specific characteristics.

#### *3.6.1.4 CE-GAN*

Consistent entrenched generative accusatorial networks (CEGAN) [16] aim to develop conditional cohort models for producing perceptually accurate outputs and arrest the entire delivery of possible multiple methods of consequences by imposing tight networks in the actual image interstellar and the hidden space. This approach circumvents the mode failure issue and generates more diversified and accurate outcomes.

As seen in Figure 3.3, The generator then reconstructs the original X2 using the participation image X1 and the secret code z. To mitigate the effect of created noise and redundancy, we encrypt the produced sample yield X2 as hidden code *z* by E. We attempt to rebuild both the latent code and picture of the ground truth to make realistic and diversified images in the goal domain.

### 3.6.2 Unsupervised Image to Image Conversion

#### *3.6.2.1 CycleGAN*

Zhu et al. [17] suggested cycleGAN for image-to-image translation challenges without paired samples in this study. They presuppose an underlying link between the domains to join "cycle consistent" loss with argumentative losses on fields X and Y. Adding two-cycle consistency losses taught a plotting from foundation domain X to goal domain Y (and vice versa).

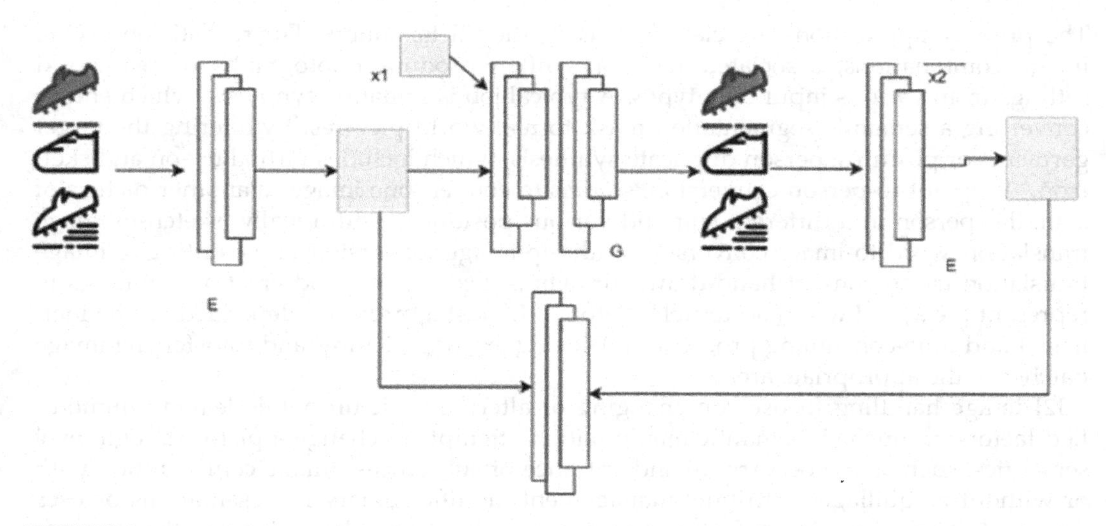


**FIGURE 3.3**
Model overview.

#### 3.6.2.2 Dugan

Z. Yi et al. [18] provide a unique dual-GAN technique for training picture interpreters using two groups of unlabeled photos from different fields. The primal GAN requires interpreting pictures from domain U to field V using its architecture, while the dual GAN studies reverse the job using its design. The closed-loop formed by primal and dual tasks enables the translation and reconstruction of pictures from either domain. As a result, the translators may be trained using a loss function that compensates for the picture reconstruction mistake.

#### 3.6.2.3 UNIT

To tackle the issue, they acquainted an unconfirmed picture with picture change organization (UNIT) in light of joined GANs and the common dormant planetary speculation, which expresses that two matching pictures from various spaces might be plotted to the specific secret portrayal in a common secret space.

#### 3.6.2.4 MUNIT

The image's satisfied code is coupled with several style illustrations taken from the goal domain to convert a picture to another domain. Each auto-hidden encoder's code comprises a satisfied code c and an elegant code s. They qualify the model using argumentative goals (dotted lines) that require the decoded pictures to be fuzzy from genuine descriptions in the goal domain and bidirectional renovation objectives (rushed lines) that need the model to reconstruct both images and latent codes.

## 3.7 Application

The primary applications are classified using the I2I taxonomy. For realistic-observing image combinations, associated I2I works often produce photographs of real-world settings from various input data types. A typical job is semantic synthesis, which entails converting a semantic segmentation mask to real-world pictures. By altering the target garments or postures, person duplicate synthesis, which includes virtual try-on and skeleton/ keypoint-to-person conversion, studies to convert one image to another picture of a similar person in a different suit and various positions. Additionally, sketch-to-image translation, word-to-image conversion, audio-to-image conversion, and painting-to-image translation try to convert hand-drawn drawings, text, audial, and creation paintings to represent the actual world accurately. Before I2I, most approaches depended on the inefficient and time-consuming process of retrieving existing pictures and transferring image patches to the appropriate area.

I2I image handling focuses on changing or altering a picture while leaving unmodified factors untouched. Semantic manipulation attempts to change a picture's high-level semantics, such as the occurrence and entrance of substances (image configuration with or without maquillage). Attribute management modifies binary representations or uses landmarks to modify picture features such as the subject's gender, hair color, the presence of spectacles, and facial expression. It also does image relabeling, gaze modification, and simulation in the wild. Additionally, image/video retargeting facilitates the transmission of consecutive material across domains while maintaining the target domain's aesthetic. Much of the I2I explore is concerned with filling in misplaced pixels, that is, image inpainting and image outpainting, although they deal with distinct types of occluded pictures. Using the concept of a human expression as an illustration, the image inpainting job generates aesthetically accurate and semantically correct outputs from input, including masked snout, mouth, and eyes.

In contrast, the outpainting image job converts a heavily blocked face image containing just a nose, aperture, and eyes. I2I contributed significantly to creative production. Historically, redrawing a picture in a specific art style required a skilled artist and some effort. In comparison, several I2I research has shown that photorealistic photos may be mechanically transformed into synthetic creations without human interaction.

Additionally, the elegance transfer job is successful when I2I techniques are used. It has two primary goals: artistic style transfer, which entails converting the participation picture to a desired imaginative elegance, such as Monet or van Gogh's, and photo accurate style transmission, which must preserve the original control structure while relocating a style. Additionally, we may use I2I for picture renovation. Image renovation aims to return a deteriorated picture to its innovative state using a deprivation model. The picture superresolution job, in particular, entails enhancing the determination of a picture, which is often skilled in using down-scaled representations of the target copy as participations. Denoising images are the process of removing noise that has been added artificially. Stunning image Dehazing and deblurring images are intended to eliminate optical imperfections from photographs shot out of attention or while the photographic camera was stirring and from photos of distant topographical or astronomical phenomena. Picture development is an independent method that utilizes heuristic processes to develop a picture pleasing to the human graphic system. I2I demonstrates its efficacy in this area, with picture colorization and duplicate quality enhancement. Colorizing a copy entails envisioning the color of each pixel based on its luminance alone. It is trained on color-disappeared pictures. The increase of image quality is focused on creating much fewer colored objects around hard limits, more precise colors, and decreased noise in smooth shadows.

Additionally, this module teaches students how to merge multi-focus images using I2I techniques. Additionally, we note that I2I algorithms use two distinct kinds of data for specific tasks: isolated sensing imaging for animal habitat investigation and building abstraction medical imaging for illness analysis dosage calculation and surgical training phantom improvement. Additionally, I2I approaches can be applied to other graphic tasks, such as transmission learning for strengthening education, image registering, domain revision, person re-identification, image segmentation, facial geometry renovation, 3D poses approximation, neural conversation head generation, and hand gesture-to-gesture conversion.

## 3.8 Conclusion

GANs have made significant progress in recent years for cross-domain image-to-image translation, and unsupervised learning techniques may now compete with supervised approaches in terms of performance. Due to the high cost or impossibility of gathering paired training data for various applications, there will be more impetus in the future for unsupervised techniques.

Additionally, we may categorize the models above as unimodal or multimodal. While UNIT, CycleGAN, and cGAN are unimodal procedures. Multimodal techniques can reduce the model collapse issue and provide diversified and realistic visuals. As a result, multimodal and unsupervised methods to picture to image translation may become the norm in the future.

## References

[1] N. Komodakis and G. Tziritas, Image completion using efficient belief propagation via priority scheduling and dynamic pruning, IEEE Trans. Image Process. 2007, 16(11): 2649–2661.

[2] I. J. Goodfellow, J. Pouget-Abadie, M. Mirza, B. Xu, D. Warde-Farley, S. Ozair, A. Courville, and Y. Bengio, Generative adversarial nets, in Proc. 27th Int. Conf. Neural Information Processing Systems, Montreal, Canada, 2014, pp. 2672–2680

[3] K. Wang, G. Chao, Y. Duan, Y. Lin, X. Zheng, F. Y. Wang, Generative adversarial networks: introduction and outlook. IEEE/CAA J Autom Sin. 2017, 4:588–98. doi: 10.1109/JAS.2017.7510583

[4] H. Emami, M. M. Aliabadi, M. Dong, R. B. Chinnam, SPA-GAN: spatial attention GAN for image-to-image translation, Computer Vision and Pattern Recognition, Cornell University, Dec 2020, pp. 391–401.

[5] E. Denton, S. Chintala, A. Szlam, and R. Fergus, Deep generative image models using a laplacian pyramid of adversarial networks. In Advances in Neural Information Processing Systems (NIPS). 2015, pp. 1486–1494.

[6] K. Gregor, I. Danihelka, A. Graves, D. J. Rezende, and D. Wierstra, DRAW: A recurrent neural network for image generation. In International Conference on Machine Learning (ICML). 2015, pp. 1462–1471.

[7] A. Dosovitskiy and T. Brox, Generating images with perceptual similarity metrics based on deep networks. In Advances in Neural Information Processing Systems (NIPS). 2016, pp. 658–666.

[8] A. Radford, L. Metz, and S. Chintala, Unsupervised representation learning with deep convolutional generative adversarial networks. 2015, arXiv preprint arXiv:1511.06434.

[9] H. Zhang, T. Xu, H. Li, S. Zhang, X. Huang, X. Wang, and D. Metaxas, Stackgan: Text to photorealistic image synthesis with stacked generative adversarial networks. 2017, arXiv preprint arXiv:1612.03242.

[10] X. Chen, Y. Duan, R. Houthooft, J. Schulman, I. Sutskever, and P. Abbeel, Infogan: Interpretable representation learning by information maximizing generative adversarial nets. In Advances in Neural Information Processing Systems (NIPS). 2016, pp. 2172–2180.

[11] T. Salimans, I. Goodfellow, W. Zaremba, V. Cheung, A. Radford, and X. Chen, Improved techniques for training GANs. In Advances in Neural Information Processing Systems (NIPS). 2016, pp. 2234–2242.

[12] M. Arjovsky, S. Chintala, and L. Bottou, Wasserstein generative adversarial networks. In International Conference on Machine Learning (ICML). 2017, pp. 214–223.

[13] Y.-J. Cao, L.-L. Jia, Y.-X. Chen, N. Lin, C. Yang, B. Zhang, Z. Liu, X.-X. Li, and H.-H. Dai, Recent advances of generative adversarial networks in computer vision. IEEE Access. 2018, 7: 14985–15006.

[14] K. Wang, C. Gou; Y. Duan, Y. Lin, X. Zheng, and F.-Y Wang. Generative adversarial networks: Introduction and outlook. IEEE/CAA J. Autom. Sin. 2017, 4: 588–598.

[15] Z. Wang, Q. She, and T.E. Ward, Generative adversarial networks in computer vision: A survey and taxonomy. arXiv 2019, arXiv:1906.01529.
[16] J. Gui, Z. Sun; Y. Wen, D. Tao, and J. Ye, A review on generative adversarial networks: Algorithms, theory, and applications. arXiv 2020, arXiv:2001.06937
[17] P. Isola, J.-Y. Zhu, T. Zhou, and A. A. Efros, Image-to-image translation with conditional adversarial networks, in Proceedings of the IEEE conference on computer vision and pattern recognition, 2017, pp. 1125–1134.
[18] Z. Yi, H. Zhang, P. Tan, and M. Gong, Unsupervised Dual Learning for Image-to-Image Translation, https://doi.org/10.48550/arXiv.1704.02510, 2017, pp. 1704–02510.

# 4

# *Generative Adversarial Networks for Video-to-Video Translation*

**Yogini Borole and Roshani Raut**

**CONTENTS**



DOI: 10.1201/9781003203964-4

## 4.1 Introduction

The purpose of video-to-video synthesis [1] is to produce an order of photo-realistic pictures from a succession of interpretation illustrations derived after a cause 3-D scene. The depictions may, for example, be interpretation subdivision covers produced by a visuals machine while operate a vehicle in a practical situation.

The depictions might alternatively be posture charts taken from a basis audiovisual of a person dancing, with the programme creating a video of another person executing the same dance [2,8].

The video-to-video synthesis problem offers numerous intriguing practical use-cases, ranging from the production of an innovative lesson of ordinal artworks to use in mainframe visuals. The capacity to create pictures that are not only separately photo-realistic, but also fleshy, is a critical prerequisite similar video-to-video combination paradigm. In addition, the produced pictures must adhere to the symmetrical and semantic construction of the underlying 3-D scene. When we see a consistent progress in photo realism and short-range temporary solidity within the generated outcomes, we trust that one risky basic part of the tricky, long-term temporal consistency, has been largely disregarded.

As an example, while returning to the same spot in the top. You would anticipate the look of earlier encountered barriers with individuals to continue intact as you proceed from site 1-N-1. The present video-to-video synthesis approaches, like video to video or our enhanced architecture integrating video to video with SPADE , however, are incapable of producing like world reliable films (third and second rows). Our only technology is capable of producing vid that are reliable across views by using a world consistency machine (first row).

Despite employing the similar explanation contributions, a current video-to-video algorithm might output a picture which considerably dissimilar from the one produced when the automobile initially stays at the place. Existing video-to-video solutions depend on optical flow warping to construct a picture based on the previous several images generated. While such processes provide short-range temporary stability, they do not provide extended span temporary consistency. Current video-to-video replicas have no idea what they have produced previously. These methodologies, in any event, for a brief full circle in a virtual room, neglect to hold the appearances of the divider and the individual in the made video.

We aim to solve the extended span temporary consistency challenge in this research by augmenting video-to-video replicas through remembrances of previous borders. We present a unique construction that clearly ensures uniformity in the whole produced cycle by merging concepts from part stream [4] and restricted picture synthesis replicas. We do extensive tests on a several of standard data sets, comparing them to state-of-the-art methods. Measurable and visual results reveal that our strategy increases image quality and long-term temporal stability significantly. On the application side, we additionally show that our methodology can be utilised to produce recordings predictable across various perspectives, empowering synchronous multi-specialist world creation and investigation.

## 4.2 Description of Background

Only a few attempts of unconditional video synthesis have been successful, in contrast to the image domain. Existing techniques for video style transfer [5,10–17] concentrate on shifting the style of a single situation picture to the output video; however, These strategies cannot be immediately applied to our multi-domain situation vid-to-vid transmission. Text inputs have mostly been employed in conditional video creation [18–19,21,22]. We are aware of individual one previous effort on general vid-to-vid fusion [23–24].

The planned prototypical studies to synthesise vid borders led by the optical flow signal that is the truth utilising aligned input and output video pairs by learning a mapping function to interpret a visual input from a single modality to another. As a result, this approach is useless for vid-to-vid transformation across semantically dissimilar areas, that is, areas with dissimilar term utterances. The lip-reading work is related to our challenge since we employ lip-reading ways to drive the generator to generate convincing videos. Previously, this problem was addressed using character level [18–19,25–26] and word level [18] categorisation methods. Our framework's goals are inextricably linked to visual speech synthesis. This issue has primarily been addressed by posture transfer. Procedures [3,5,7,26] produce vid by pictorial language by shifting one speaker's posture (lip movement) to another provided with a target speaker and a driving vid. A separate method is language-driven face fusion [6,8,9,11,27]. In this design, the movies are based on a picture of the target speaker as well as an audio signal. The algorithms in [23] create a person's speaking facial animation using a single face image and a sound stream. Gao et al. [7] make video from noises in a similar manner, but it generates frames independently of each other and just animates the lips. The authors of a recent paper [26,29–30] employ a generative adversarial networks-based strategy to produce conversation expressions that are conditioning on both video and sound, which is suited for audiovisual organisation and differs after our arrangement.

In this chapter, we discuss the different methods and architectures used for video-to-video translation.

We describe the suggested framework for video-to-video translation with global and local temporal consistency in this section.

### 4.2.1 Objectives

To the best of our knowledge, we are the first to present a single framework for solving various video-to-video translation challenges. We use the suggested system to execute three tasks to test its effectiveness: video super resolution, video colourisation, and video segmentation. The suggested framework is sufficiently versatile, and it may employ existing video producing methods directly as the generator in our framework.

We analyse the temporal inconsistency in video-to-video translation tasks and suggest a unique two-channel discriminator to achieve global temporal consistency for the whole testing video. Furthermore, by combining the current local temporal consistency with the proposed global temporal consistency in a unified framework, our solution outperforms the single local consistency significantly.

## 4.3 Different Methods and Architectures

GAN is one of the most intriguing and promising machine learning innovations. The GAN is a period of semantic net that competes against respectively other in the way of

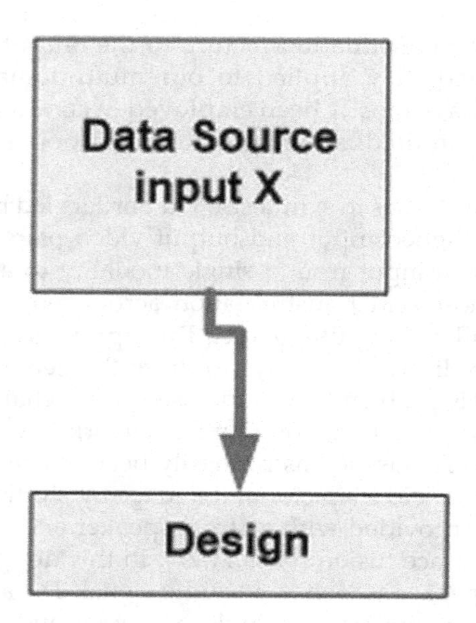


**FIGURE 4.1**
Unsupervised learning example.

a lose–lose situation. This approach learns to produce fresh information with similar measurements as the preparation set given a preparation sets. It is an unverified education activity in machine learning that entails mechanically detecting and knowledge designs in approaching information so the model might be utilised to create or yield new cases from the first dataset. Only the inputs are provided as shown in Figure 4.1 in unsupervised learning, and the purpose is to uncover stimulating designs in the information.

## 4.4 Architecture

A GAN is made up of two components: The creator and the differentiator as shown in Figure 4.2. The generator studies to produce likely data based on the training information, while the differentiator attempts to differentiate among the generator's phony information and genuine data. The generator is penalised by the discriminator for producing improbable outcomes. The creator attempts to deceive the differentiator into believing that the created imageries are actual, while the differentiator attempts to distinguish between actual and phony imageries. Casual sound is supplied into the creator, which converts it to a "false picture." The primary idea underlying generative adversarial networks is to alternately trains the creator and differentiator to be the finest they can be at creating and discerning pictures. The goal is that by enhancing one of these networks in this game-theoretic way, the other network will have to do an improved work to success the willing, which in turn increases presentation, and so on. For unsupervised tasks, Generative Adversarial Networks have shown outstanding performance.

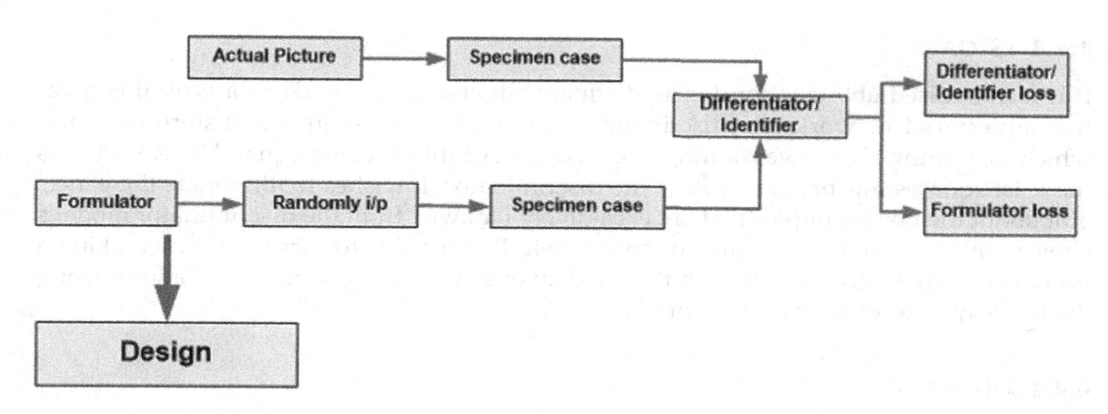


**FIGURE 4.2**
Generalised architecture of generative adversarial networks.

The following are some of the utmost frequently used GAN architectures:

A GAN is made up of two components: The creator and the differentiator as shown in Figure 4.2. The generator studies to produce likely data based on the training information, while the differentiator attempts to differentiate among the generator's phony information and genuine data.

### 4.4.1 Cycle GAN

It is a method for training picture-to-picture translation models automatically without the need of corresponding instances. A horse to a zebra translation is an example of picture-to-image translation. Using a cycle consistency loss, cycle GAN trains without the need for paired data. It can translate from one field to another without needing one-to-one planning between the base and objective domains. The goal of the picture-to-picture translation problem is to learn the planning between an input picture and an output picture with a training set of matched picture pairings.

Cycle GAN learns a mapping G:X → Y and F: Y → X for two domains X and Y.

Cycle GAN is divided into three processing segments: encoding, transforming, and decoding.

Cycle GAN can be useful when colour transformation is required.

### 4.4.2 Style GAN

Style GAN is a unique GAN developed by Nvidia researchers. It is a generative adversarial network architecture extension that proposes significant changes to the creator prototypical, for example, the utilisation of an arranging net to design focuses in dormant space to a halfway inactive area, the utilisation of the moderate inactive area to control style at every point in the creator model, and the presentation of commotion as a wellspring of variety at every point in the creator model. The resulting model can not only create very accurate superb images of appearances, but it also considers command over the style of the created image at various locations of detail by modifying the style routes and clamour.

### 4.4.3 LS-GAN

It is a formalised abbreviation for least square adversarial network. In a typical generative adversarial network, the discriminator employs a cross-entropy misfortune work, which may bring about evaporating slope issues. All things being equal, LSGAN utilises the least-squares misfortune work as the discriminator. It wishes to illuminate the generator about the fake examples that are excessively far away from the discriminator model's choice limit to be named genuine or counterfeit. It is feasible to carry out it by making a humble change in accordance with the yield layer of the discriminator layer and utilising the least-squares, or misfortune work.

### 4.4.4 Disco GAN

Given a photograph in domain A, it creates images of products in domain B. It moves components from one image to the next.

Disco GAN is powered by two creators. The initial creator converts the input picture from field A to field B. The picture is reconstructed from domain B to domain A using the second generator.

Cycle GAN and disco GAN have certain similarities. Cycle GAN includes additional hyperparameters for adjusting the commitment of reproduction/cycle-consistency misfortune to the complete misfortune work. Disco GAN can be used to transfer ornamentation from one style article, such as a purse, to another, such as a pair of shoes.

### 4.4.5 Mo-cycle GAN

Motion-guided cycle generative adversarial network construction is used for integrating gesture approximation into unpaired video interpreter, investigating both appearance design and worldly progression in video interpretation. Figure 4.3 depicts the whole construction of mo-cycle GAN. This section begins by expanding on the terminology and issue detailing of unpaired video-to-video interpretation, followed by a quick overview of cycle GAN with spatial limitation. Then, at that point, to build up the transient coherence, two kinds of movement directed worldly requirements are presented: movement cycle consistency and movement interpretation.

Thusly, both the visual examine each edge and the movement between progressive casings are destined to be reasonable and steady across transformation. Finally, the enhancement approach for the preparation and deduction stages is portrayed.

### 4.4.6 Different GANs for Video Synthesis (Fixed Length)

GANs for video are roughly classified into two types: models that generate static span movies and replicas that permit for variable span audiovisual combination. The former generates 3D (thickness, altitude, and period) video chunks. Because of the limitations of their designs, such models can only output films of the same duration. This category includes common instances such as VGAN [9] and TGAN [10].

VGAN [9] is among the primary endeavours to produce transiently rational video tests through generative adversarial network design. Audiovisual signals are deteriorated into two distinct categories: a stationary "foundation" category and a moving "forefront" stream. Utilising a lot of unspecified audiovisual, the author proposes V GAN, a

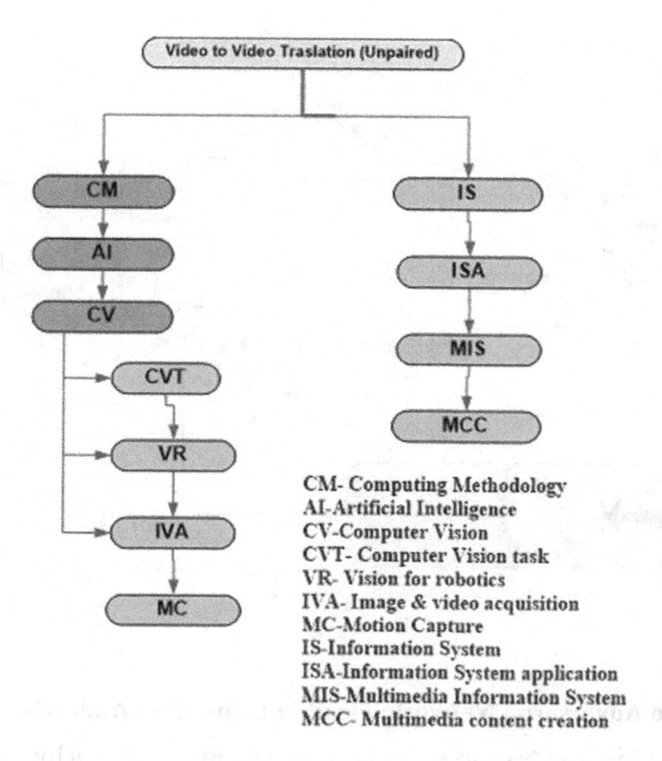


**FIGURE 4.3**
Depiction of the whole construction of mo-cycle GAN.

GAN-based genuine reproductive prototypical that figures out how to show practical frontal area movements and a stationary foundation.

As the principal endeavour to convey the achievement of generative adversarial network in the picture space to the video space, VGAN has exhibited a doable way of learning movement elements. Trials have likewise recommended VGAN can learn highlights for human activity order without any oversight. Notwithstanding, the unequivocal partition of the closer view and the foundation makes VGAN unfit to learn more confounded recordings. The supposition of a fixed foundation significantly confines the handiness of VGAN. The generally basic and crude network constructions and misfortune capacities forestall the model to deliver recordings of higher visual quality. All the more critically, VGAN's design doesn't consider producing recordings with subjective lengths.

### 4.4.7 TGAN

TGAN produces edges and fleeting elements independently. Given a commotion vector z0, the worldly generator G0 correlates it to a progression of idle codes, individually addresses an inactive point in the picture area. The picture creator G1 takes in both z0 and produced idle codes to deliver video outlines. The structure of TGAN's engineering is shown in Figure 4.4.

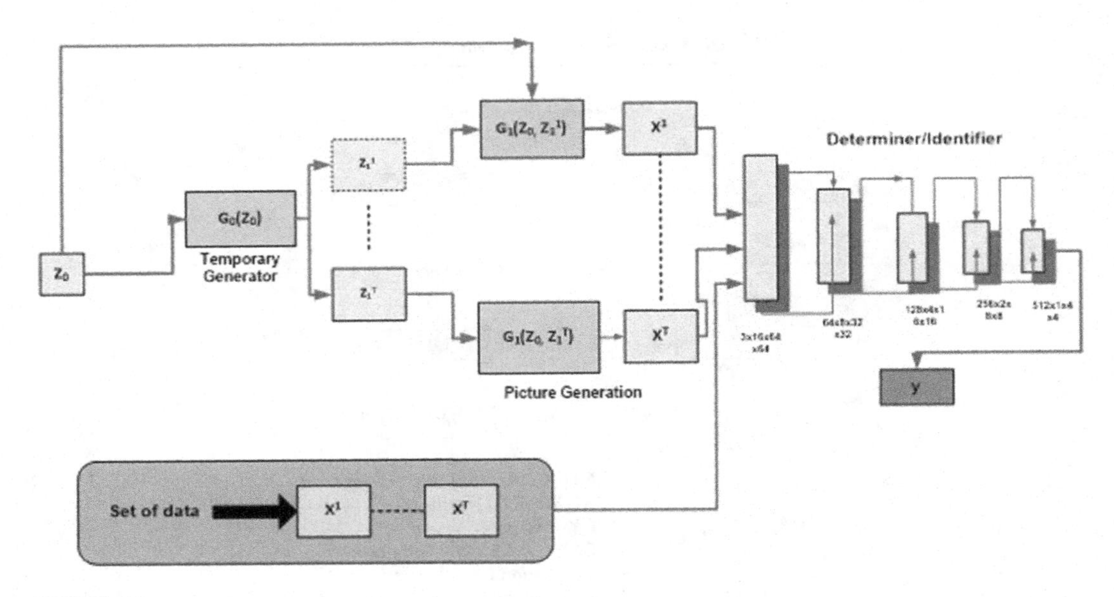


FIGURE 4.4
T GAN architecture.

### 4.4.8 Generative Adversarial Network: Flexible Dimension Audiovisual Combination

The other sort of video GANs successively produce video outlines by a repetitive neural organisation. The repetitive neural network permits the model to produce a subjective amount of edges.

MoCo GAN [11] and DVD GAN [12] have a place with this classification.

#### *4.4.8.1 MoCo GAN*

Both V GAN [9] and T GAN [10] accept an audiovisual test by means of a fact in an idle area. Their models are intended to change a fact in the inert space over to an audiovisual cut. Isola et al. [11] contend that such a presumption pointlessly makes the issue more complicated.

Above all else, recordings with a similar activity yet unique speed are viewed as various focuses in this methodology. All the more significantly, this methodology dismisses the way that recordings have various lengths. Isola et al. [11], on the other hand, anticipate that a video will be created by travelling numerous foci in an inactive area of photos. Under this notion, records of varying speeds and lengths are compared to directions in inert space intersected by varying speeds and durations. The new approach allows for the decomposition of the inert space of images into two subspaces: the substance subspace and the movement subspace. Based on the knowledge, an unqualified video generative adversarial network (MoCo GAN) is planned. Rather than straightforwardly planning an inactive vector to a video, MoCo GAN produces transiently intelligible casings consecutively. Each edge's inactive code consists of a "content" code that is randomly tested from Gaussian conveyance and is fixed during the whole video cut and a "movement" code that is formed by a learnt repeated neural organisation. Fixing the "content" portrayal while evolving the "movement" portrayal forces the model to separate the two subspaces and provide realistic recordings with trustworthy items and understandable motions. The

MoCo GAN comprises of a one-layer GRU network RM, a picture generator organisation GI, a picture discriminator network DI, and a video discriminator network DV. During preparing, the substance code ZC is inspected once and fixed for the whole video. A progression of arbitrarily inspected vectors $[\mathcal{E}^{(1)} \ldots\ldots \mathcal{E}^{(k)}]$ are input into the RM and encoded to $K$ movement codes $[Z^{(1)} M, \ldots, Z(K) M]$ one at a stage. The picture generator GI guides connected ZC and $Z^{(k)} M$ to the $k$th outline xe($k$). DI is a standard two-dimensional convolutional neural network engineering and just evaluates the spatial nature of each single casing. DV contains spatio-transient convolutional neural network layers with the goal that it can take in a fixed-length video clasp and let know if it is manufactured or legitimate.

In spite of the fact that MoCo GAN has outperformed its archetypes like V GAN and T GAN in wording of quantitative assessment measurements on a few datasets, we think that it is imperative to call attention to the following issues:

1. The disintegration of visual signs in a video to a substance part and a movement part misrepresents the issue by expecting all recordings should have a fixed what's more, clear cut article a still foundation. This supposition might assist the MoCo GAN with excelling on exceptionally controlled datasets like MUG Facial Expression Database and Weizmann Action Information Base [19]. With regards to more mindboggling and different datasets like UCF 101, MoCo GAN experiences issues introducing numerous articles, deformable items or camera panning in yield tests. As indicated by our investigations, MoCo GAN might accomplish a somewhat preferable Inception Score over T GAN and V GAN on UCF 101 [20], the visual nature of the manufactured examples is still exceptionally terrible.
2. In any event, for those datasets that MoCo GAN performs better on, the enhancements could be the aftereffect of generative adversarial network remembrance. Datasets utilised in the investigations are tiny. But UCF101, the biggest dataset utilised in preparing contains just 4,500 clasps, almost multiple times less than Kinetics-400 [11,13].

#### *4.4.8.2 DVD GAN*

The achievement of Big GAN has been extended to the video realm. Double video discriminator generative adversarial network (DVD GAN) [22], a class-restricted GAN construction created on Big GAN, is intended for video aggregation and video expectation. In our work, we focus on the video amalgamation component.

Many audiovisual generative adversarial network models [9,10,11] that came before DVD GAN mostly dealt with small and basic video datasets, such as the Moving MNIST dataset and the Weizmann Action database [19]. DVD GAN can produce high consistency 256 recordings with 48 casings and achieve best in class results on both the UCF 101 [20] and Kinetics 600 [22]. Because of the strong learning ability of its neural organisations, DVD GAN does not indisputably segregate frontal area from foundation as in [11] or construct new casings via stream distorting and the visualisation network as in [25,28–32]. DVD GAN has two discriminators: a spatial discriminator DS and a transient discriminator DT. The use of convolutional GRU and the presentation of separable attention module, a spatio-transient augmentation of the self-consideration module provided by [20] for picture amalgamation, are the main improvements of DVD GAN. Kim et al. [12] use the multiplicative convolutional GRU for tests with less than 48 casings. To save memory and computational costs, we use the conventional convolutional GRU in our training. The majority of Big GANs successful low-level compositional judgements are obtained via DVD GAN.

#### *4.4.8.3 Methods and Tools for GAN*

For a couple of years at this point, GANs have been effectively utilised for high-constancy regular picture amalgamation, information expansion, and that's only the tip of the iceberg. From making photograph reasonable talking head models to pictures uncannily looking like human faces, GANs have taken tremendous steps of late. Underneath, we have curated a rundown of the best 10 apparatuses for GAN.

#### *4.4.8.4 GAN Lab*

GAN Lab is an intelligent, visual experimentation instrument for GANs. With this instrument, you can intelligently prepare GAN models for 2D information appropriations just as envision their internal operations. For execution, GAN Lab utilises TensorFlow.js, an in-program GPU-sped up profound learning library. GAN Labs representation abilities can be utilised to figure out how the generator of a model gradually updates to work on itself to give counterfeit examples that are progressively more reasonable. A portion of its highlights are:

- Intelligent hyperparameter change
- Client characterised information dissemination
- Slow-movement mode
- Manual bit by bit execution.

### 4.4.9 Hyper GAN

Hyper GAN is a composable GAN structure that incorporates API and UI. Hyper GAN constructs generative antagonistic organisations in PyTorch and makes them simple to prepare and share. Hyper GAN is intended to help custom exploration also. With this structure, one can without much of a stretch supplant any piece of the GAN with the json document, or simply make another GAN through and through. As of now, hyper GAN is in pre-delivery and open beta stage.

### 4.4.10 Imaginaire

Created by NVIDIA, Imaginaries is a PyTorch-based GAN library that incorporates every one of the streamlined executions of different pictures and video amalgamation projects formed by Nvidia into one. It is a multi-reason library with various usefulness, from picture handling to video interpretation and generative style move. Delivered under the Nvidia programming permit, Imaginaire contains six calculations that help picture-to-picture interpretation: pix2pixHD, SPADE, FUNIT, UNIT, MUNIT, and COCO-FUNIT.

### 4.4.11 GAN Tool Compartment

The GAN tool compartment by IBM utilises an incredibly adaptable, no-code way to deal with carry out the well-known GAN models. By giving the subtleties of a GAN model as order line contentions or in an instinctive config document, the code could be effortlessly produced for preparing the GAN model. A portion of the benefits of the GAN tool compartment are the following.

It is an exceptionally modularised portrayal of GAN model for simple blend and-match of parts across structures. It gives a theoretical portrayal of GAN engineering to give multi-library support. It is a coding free method of planning GAN models.

### 4.4.12 Mimicry

Mimicry is a well-known and lightweight library in PyTorch pointed towards the reproducibility of GAN research. Mimicry offers Tensor Board help for haphazardly produced pictures for actually taking a look at variety, misfortune, and likelihood bends for observing GAN preparing. A portion of its highlights are as follows:

- Normalised the executions of GAN models that intently duplicate announced scores
- Benchmark scores of GANs prepared and assessed under similar conditions
- A structure to zero in on execution of GANs without revamping the vast majority of GAN preparing code.

### 4.4.13 Pygan

Pygan is a well-known library written in Python. The library is utilised to carry out models like GANs, adversarial auto-encoders (AAEs), and conditional GANs just as energy-based GAN (EBGAN). The library makes it conceivable to plan the generative models dependent on the statistical AI issues corresponding to GANs, conditional GANs, AAEs, and EBGAN to rehearse calculation plan for semi-administered learning.

### 4.4.14 Studio GAN

Studio GAN is a famous library in PyTorch for carrying out delegate GANs for restrictive just as unrestricted picture age. The library incorporates a few fascinating highlights, for example,

- Broad GAN executions for PyTorch
- Better execution and lower memory utilisation than unique executions
- Far-reaching benchmark of GANs utilising CIFAR10, Tiny ImageNet, and ImageNet datasets.

### 4.4.15 Torch GAN

Torch GAN is a well-known PyTorch-based structure, utilised for planning and creating GANs. This structure has been explicitly intended to give building squares to well-known GANs. It additionally permits customisation for state-of-the-art research.

This structure has various highlights, for example,

- Permits to give a shot to well-known GAN models on your dataset.
- Permits to connect your new Loss Function, new Architecture, and so forth with the conventional ones.
- Helps in consistently envisioning the preparation with an assortment of classification backends.

### 4.4.16 TF-GAN

Tensor flow GAN is a lightweight library for preparing just as assessing GAN. Profoundly, normal GAN activities and standardisation strategies, misfortunes, and punishments, among others.

### 4.4.17 Ve GANs

Ve GANs is a Python library with different current generative adversarial network in PyTorch. All the more explicitly, the library can without much of a stretch train different current generative adversarial network in PyTorch. The library is intended for clients who need to utilise existing GAN preparing procedures with their own creators/differentiators. The client gives differentiator and creator organisations, and the collection deals with preparing them in a chose GAN setting.

## 4.5 Conclusions

There are large number of articles on GANs and a large number of named GANs, that is, models with a characterised term that frequently incorporates GAN. There can be a few challenges while making another engineering utilising the GAN design. It is realised that the preparation GANs is exceptionally temperamental. Meta parameter streamlining is hard to track down the right harmony between learning rates to proficiently prepare the organisation. In every one of the structures, generators are generally utilised for preparing the data edited by DVD GAN.

## References

1. Badrinarayanan, V.; Kendall, A.; and Cipolla, R. 2017. SegNet: A deep convolutional encoder-decoder architecture for image segmentation. IEEE Transactions on Pattern Analysis and Machine Intelligence 39(12): 2481–2495.
2. Huang, S.; Lin, C.; Chen, S.; Wu, Y.; Hsu, P.; and Lai, S. 2018. AugGAN: Cross domain adaptation with GAN-based data augmentation. In Proceedings of the European Conference on Computer Vision (ECCV).
3. Chang, H.; Lu, J.; Yu, F.; and Finkelstein, A. 2018. PairedCycleGAN: Asymmetric style transfer for applying and removing makeup. In Proceedings of the IEEE conference on computer vision and pattern recognition.
4. Chen, M. H.; Kira, Z.; AlRegib, G.; Woo, J.; Chen, R.; and Zheng, J. 2019. Temporal attentive alignment for large-scale video domain adaptation. In Proceedings of the IEEE/CVF International Conference on Computer Vision.
5. Chen, Y.; Lai, Y.; and Liu, Y. 2018. Cartoon GAN: Generative adversarial networks for photo cartoonization. In CVPR.
6. Denton, E.; Chintala, S.; Szlam, A.; and Fergus, R. 2015. Deep generative image models using a Laplacian pyramid of adversarial networks. *Advances in Neural Information Processing Systems* 28: 1–9.

7. Gao, Y.; Liu, Y.; Wang, Y.; Shi, Z.; and Yu, J. in press, 2018. A universal intensity standardization method based on a many-toone weak-paired cycle generative adversarial network for magnetic resonance images. *IEEE Transactions on Medical Imaging* 38.9: 2059–2069.
8. Zhang, B.; He, M.; Liao, J.; Sander, P.V.; Yuan, L.; Bermak, A.; and Chen, D. 2019. Deep exemplar-based video colorization. In Proceedings of the IEEE Conference on Computer Vision and Pattern Recognition. pp. 8052–8061.
9. Eigen, D., and Fergus, R. 2015. Predicting depth, surface normals and semantic labels with a common multi-scale convolutional architecture. In Proceedings of the IEEE International Conference on Computer Vision. pp. 2650–2658.
10. Huang, S.; Lin, C.; Chen, S.; Wu, Y.; Hsu, P.; and Lai, S. 2018. AugGAN: Cross domain adaptation with GAN-based data augmentation. In Proceedings of the European Conference on Computer Vision (ECCV).
11. Isola, P.; Zhu, J. Y.; Zhou, T.; and Efros, A. A. 2017. Image-to-image translation with conditional adversarial networks. In Proceedings of the IEEE conference on computer vision and pattern recognition.
12. Kim, T.; Cha, M.; Kim, H.; Lee, J.; and Kim, J. 2017. Learning to discover cross-domain relations with generative adversarial networks. In International conference on machine learning. PMLR.
13. Kingma, D., and Ba, J. 2014. Adam: A method for stochastic optimization. arXiv preprint arXiv:1412.6980.
14. Ledig, C.; Theis, L.; Huszar, F.; Caballero, J.; and Shi, W. 2017. Photo-realistic single image super-resolution using a generative adversarial network. In Proceedings of the IEEE conference on computer vision and pattern recognition.
15. Lee, H. Y.; Tseng, H. Y.; and M. K. Singh, J. B. H.; and Yang, M. H. 2018. Diverse image-to-image translation via disentangled representations. In Proceedings of the European conference on computer vision (ECCV).
16. Jiang, Liming, et al. 2020. "Tsit: A simple and versatile framework for image-to-image translation." European Conference on Computer Vision. Springer, Cham.
17. Ma, S.; Fu, J.; Chen, C. W.; and Mei, T. 2018. DA-GAN: Instancelevel image translation by deep attention generative adversarial networks. In Proceedings of the IEEE conference on computer vision and pattern recognition. pp. 5657–5666.
18. Wang, Y.; Wu, C.; Herranz, L.; van de Weijer, J.;. Gonzalez-Garcia, A.; and Raducanu, B. 2018. Transferring GANs: generating images from limited data, in Proceedings of the European Conference on Computer Vision (ECCV). pp. 218–234.
19. Wang, Z.; Lim, G.; Ng, W.Y.; Keane, P.A.; Campbell, J.P.; Tan, G.S.W.; Schmetterer, L.; Wong, T.Y.; Liu, Y.; and Ting, D.S.W. 2021. Curr Opin Ophthalmol. Sep 1;32(5):459–467. doi: 10.1097/ICU.0000000000000794.PMID: 34324454
20. Zhao, L.; Peng, X.; Tian, Y.; Kapadia, M.; and Metaxas, D. N. 2020. Towards image-To-video translation: A structure-aware approach via multi-stage generative adversarial networks. International Journal of Computer Vision 128(10–11): 2514–2533. https://doi.org/10.1007/s11263-020-01328-9.
21. Vazquez, D.; Bernal, J.; S ′ anchez, F. J.; Fern ′ andez-Esparrach, G.; ′ Lopez, A. M.; Romero, A.; Drozdzal, M.; and Courville, A. 2017. ′ A benchmark for endoluminal scene segmentation of colonoscopy images. Journal of healthcare engineering 2017: 1–10.
22. Wang, P.; Chen, P.; Yuan, Y.; Liu, D.; Huang, Z.; Hou, X.; and Cottrell, G. 2018a. Understanding convolution for semantic segmentation. In WACV.
23. Wang, T.; Liu, M.; Zhu, J.; Tao, A.; Kautz, J.; and Catanzaro, B. 2018b. High-resolution image synthesis and semantic manipulation with conditional GANs. In Proceedings of the IEEE conference on computer vision and pattern recognition.
24. Wang, T. C.; Liu, M. Y.; Zhu, J. Y.; Liu, G.; Tao, A.; Kautz, J.; and Catanzaro, B. 2018c. Video-to-video synthesis. In *arXiv preprint arXiv:1808.06601.* https://doi.org/10.48550/arXiv.1808.06601.

25. Olszewski, K.; Li, Z.; Yang, C.; Zhou, Y.; Yu, R.; Huang, Z.; Xiang, S.; Saito, S.; Kohli, P.; and Li, H. 2017. Realistic dynamic facial textures from a single image using GANs. In IEEE international conference on computer vision (ICCV).
26. Bin, Z.; Li, X.; and Lu, X. 2017. Hierarchical Recurrent neural network for video summarization. In Proceedings of the 25th ACM international conference on Multimedia.
27. Zhang, Z.; Yang, L.; and Zheng, Y. 2018. Translating and segmenting multimodal medical volumes with cycle- and shape consistency generative adversarial network. In Proceedings of the IEEE conference on computer vision and pattern Recognition.
28. Deng, W.; Zheng, L.; Ye, Q.; Kang, G.; Yang, Y.; and Jiao, J. 2018. Image-image domain adaptation with preserved self-similarity and domain-dissimilarity for person re-identification. In Proceedings of the IEEE conference on computer vision and pattern recognition.
29. Ronneberger, O.; Fischer, P.; and Brox, T. 2015. U-Net: Convolutional networks for biomedical image segmentation. In International Conference on Medical image computing and computer-assisted intervention. Springer, Cham.
30. Wei, L.; Zhang, S.; Gao, W.; and Tian, Q. 2018. Person transfer GAN to bridge domain gap for person re-identification. In Proceedings of the IEEE conference on computer vision and pattern recognition.
31. Zanjani, F. G.; Zinger, S.; Bejnordi, B. E.; van der Laak, J. A. W. M.; and de With, P. H. N. 2018. Stain normalization of histopathology images using generative adversarial networks. In 2018 IEEE 15th International symposium on biomedical imaging (ISBI 2018). IEEE.
32. Zhu, J.; Park, T.; Isola, P.; and Efros, A. A. 2017. Unpaired image-to-image translation using cycle-consistent adversarial networks. In Proceedings of the IEEE international conference on computer vision..

# 5

# *Security Issues in Generative Adversarial Networks*

**Atul B. Kathole, Kapil N. Vhatkar, Roshani Raut, Sonali D. Patil, and Anuja Jadhav**

**CONTENTS**



## 5.1 Introduction

Attacks against digital information, whether it is at relaxation or in movement, continue to increase in frequency, intensity, and significance. The Net evolved from an intellectual curiosity to a critical feature of virtually everyone who utilizes or is impacted by contemporary technology in their everyday lives. Hackers rapidly realized the importance of the data on the Net and society's reliance on network connection and communication. The next evolution stage in modern networking includes manufacturing and detection devices, dubbed the Internet of Things (IoT). By 2030, it is predicted that 50 billion gadgets will be put to the previously dense system, creating many additional entrance points and possibilities for manipulation [1]. Cyber-attacks are a continuous threat to the numerary life of people. As Dutta et al. point out in [2], gadgets and systems are

DOI: 10.1201/9781003203964-5

susceptible to catastrophic charge that negatively impact consumers' everyday lives [3]. Complex intrusions are being developed using sophisticated algorithms to bypass recognition systems that have not dealt with the increasing difficulty of threats. Attackers may keep, analyze, and sell data about the operator's digital lifespan. Frequently, these hacked structures are coupled with other cooperated facilities to accomplish more goals; snowballing is a process. These risks, along with a deficiency in effective detection mechanisms, expose sensitive data. Preventing these attacks on systems and networks is a critical cybersecurity objective.

Although safety systems are becoming more resistant and secure due to training for a broader range of charges, many attacks remain unaccounted for. Additionally, charges are often found. Thus, one could wonder how a defense might protect themselves from an assault if they are unaware of its existence. As a consequence, recognition systems are employed to safeguard systems against known threats. The following are some instances [4] of charge on structures that are insufficiently protected, which prompted this review:

- Weather Network Ransomware: In April 2019, amid a significant weather event in the southern United States, a ransomware assault rendered the cable station inoperable for extra than an hour [5]. This led to the possibility of possessions and life lost due to the absence of critical data.
- Capital One Fissure: In July 2019, hackers stole personal identifying information from thousands of credit card applicants, including their birthdays and Social Security numbers [6].
- Texas Ransomware: In August 2019, a ransomware assault targeted the computers of 22 small communities in Texas. As a consequence, the government was prevented from issuing birth and death credentials [7].

With the introduction of machine learning (ML), investigators started using ML's capabilities to enhance their safety systems. ML is a technique in which a computer program improves its performance over time depending on previous experience. The more time provided for data-driven training, the more data it can collect to adjust to stated objectives. The following picture illustrates the fundamental GAN architecture.

GANs seem to be an excellent method for training a neural network to simulate a known assault. GANs may generate synthetic data based on their training experience with real-world data; consequently, various potential charges, including previously unknown ones, are investigated utilizing this produced data that a user would not usually provide during training. This enables the GANs approach to be very effective in system security where new and unexpected data threats are being created regularly.

By using GANs, a designer may isolate and anticipate future charge, allowing for the development of improved security measures to ward off such attacks before the attackers ever imagine them [8]. Additionally, hackers may utilize this technique to generate new risks. The drive of this study is to observe the security offered and the threats modeled using GANs. The use of GANs for safety is a fascinating area of study. A thorough review of the literature in this field revealed that no prior surveys had addressed the present work using GANs in cybersecurity. The purpose of this chapter is to provide an overview of how GANs have been developed and used to improve security actions. Particular focus is put on the use of GANs to detect threats and charge against systems.

## 5.2 Motivation

One of the ongoing struggles for cybersecurity is to learn to protect yourself from newer, more sophisticated attacks you've never seen before. Cyber-attacks are a continuous threat to the numerary life of people. As Dutta et al. point out, gadgets and systems are susceptible to catastrophic charge that negatively impact consumers' everyday lives. Complex intrusions are being developed using sophisticated algorithms to bypass recognition systems that have not dealt with the increasing difficulty of threats. Attackers may keep, analyze, and sell data about the operator's digital lifespan. Frequently, these hacked structures are coupled with other cooperated facilities to accomplish more goals; snowballing is a process. These risks, along with a deficiency in effective detection mechanisms, expose sensitive data. Preventing these attacks on systems and networks is a critical cybersecurity objective. However, if you take the time to understand GAN, you can protect your data for the future.

### 5.2.1 Objectives

Following are some basic objectives on which we are going to focus through this chapter:

- Working of GAN network
- How to detect the unknow activity in network
- To secure the network from unauthorized access

The following sections summarize the study: Section 5.2 discusses related work on GANs, their architecture, and the notion of fundamental system security. Section 5.3 covers potential GAN charge; Section 5.4 concludes.

## 5.3 Related Work

The following section gives a detailed overview of GANs related work and security issues in the current network.

### 5.3.1 Generative Adversarial Network

In 2014, Ian Goodfellow proposed the concept of GANs [8]. As shown in Figure 5.1, it is a two-network structure comprised of a producer and a discriminator [9]. The network trains adversarially, with the producer model G attempting to replicate the information distribution using an arbitrary noise vector given to the discriminator model D. D [10,11] evaluates the model and returns a possibility value indicating whether the information originated in G or a genuine information collection [12]. The output of D is used to train G. G's objective is to increase the likelihood that D believes the data came from the actual dataset [13,14].

### 5.3.2 Overview of Security

Cybersecurity is critical to information technology [15] because it provides safeguards against different types of digital charge on systems, networks, and programs [16]. The

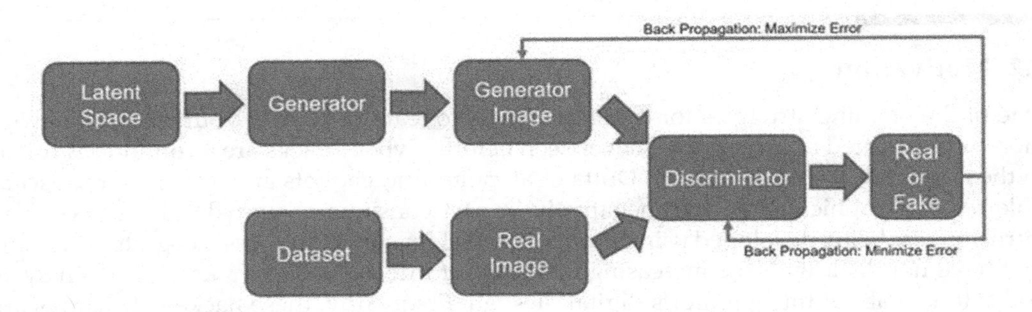


**FIGURE 5.1**
The architecture of GAN.

modern world is inextricably linked [17,18]. By 2030, it is estimated that the average individual would possess fifteen connected gadgets [19].

Consequently, there are a plethora of targets with very disparate degrees of protection, which complicates the task of securing all of those operators effectively [20]. With the growing number of linked gadgets and the inventiveness of assailants, cybersecurity risks are continuously changing, and the numeral of potential charge is growing [21,22]. Artificial Intelligence (AI) has an important use in the field of security [23]. While developments in AI and machine learning procedures reduce the work needed to develop more secure systems, they also generate and expose new methods for compromising ostensibly safe structures. GANs are relatively new knowhow that makes both good and bad modifications to an area of system design that is often overlooked - security.

### 5.3.3 GANs in Safety

The following section gives the detailed overview of GANS network safety and how effectively we can work on it.

#### *5.3.3.1 Obscuring Delicate Information*

Businesses and governmental organizations often hold sensitive data that researchers are not permitted to access. For instance, healthcare institutions store a great deal of sensitive patient data. Banks additionally protect their customers' statements and financial data. If protected data could be shared with researchers or analysts, it would undoubtedly provide critical insights and aid future studies. WWT artificial intelligence research [24] demonstrates that an appropriately skilled GAN may generate new information representing the original. As a result, the actual data may be protected, but the data produced by the GANs is probably retains similar patterns and visions as to the original information. The assembled information has a high degree of similarity to the features of the original information set. This information may be examined more to ensure the original information's safety. That study assessed the produced information's similarity to the original data and the viability of developing an analytical model utilizing the generated data. Mirjalili et al. [25] also stated that a computerized analysis might be used for age- or gender-based outlining, undermining the use of biostatistics in various claims [26]. They presented an auto-encoder based on an improved GAN that transforms a participation face picture into one that can be utilized for facial acknowledgment but not gender categorization.

### 5.3.4 Cyber Interruption and Malware Detection

Cyber interruptions are used to disrupt and compromise computers by breaching their safety system or creating an unsuitable operating environment. Cyber invasions have a variety of consequences, including illegal publishing of information, information manipulation, and data loss. An interruption detection system (IDS) displays the system and notifies the user if it detects any harmful activity. Using statistical learning and NNs, researchers have been examining various IDSs [27]. Chen et al. [28] presented that a GAN-based prototypical might be a highly effective option for executing an IDS in their study. By knowledge of the characteristics of regular information, GANs are utilized in intrusion detection.

Chen et al. presented a GAN-based prototypical with a revised loss role and several intermediary sheets to get reasonable choices from the discriminator in their study. Scientists have also used GANs to identify malware. Malware is a malicious package that is developed with the express purpose of causing harm to a computer system, and malware is not always used in invasions. Anderson et al. [29] established adversarial-tuned field creation. Also, they proved that supplementing the exercise set with produced accusatorial instances enables the classifier to identify more malware relations than using previous methods. Burks et al. [30] conducted a comparison analysis of GANs and variational autoencoder (VAE) representations and found that the efficiency of malware recognition increased when GANs were used. Additionally, many academics have suggested novel defensive methods and algorithms against various charge based on GANs.

Samangouei et al. [31] developed Defense-GAN, a generative adversarial network skilled to simulate the delivery of unperturbed pictures. Their suggested GAN may be used to defend against a variety of occurrence techniques and also improve on already utilized defensive measures [32]. The majority of currently deployed defensive methods are model-specific [33,34]. Using GANs' generating capabilities, Defense-GAN may be employed with any classifier and on any assault [35].

### 5.3.5 Security Examination

Cyber-physical production schemes (CPPSs) are a new security infrastructure that integrates the electronic and physical security levels. Such a system is precious economically and socially. The concept of CPPS refers to integrating computing, networking, and bodily processes in independent and linked subsystems [47]. Cross-domain security examination of cyber and human systems is a critical area of study. Chhetri et al. [48] proposed the GAN-Sec scheme, a security analysis model based on the GAN algorithm. GAN-Sec performs safety analysis by examining a system's signal and energy flows. GANs may assist in determining if the major security apparatuses (confidentiality, accessibility, and integrity) are fulfilled. The authors used an additive manufacturing structure as a case study to demonstrate GAN-Sec's usefulness shown in Table 5.1. The findings indicate that the suggested model can analyze security breaches caused by side-channel occurrences [49]. Additionally, GAN-Sec can assist in assessing the efficacy of a model for detecting integrity and availability attacks [50,51].

**TABLE 5.1**

Analysis of GAN Security Issues

| Purpose of Work | References | GAN Type Used in Reference Paper | Challenges |
|---|---|---|---|
| Obscuring delicate information | WWT AI [24] | Vanilla GAN [8] | Observe the security issues |
| | Mirjalili et al. [25] | A modified version of GAN | |
| Cyber interruption and malware recognition | Chen et al. [28] | BiGAN [50] | Work on cyber and malware attack but not focus on how it can be detected |
| | Anderson et al. [29] | Vanilla GAN | |
| | Burks et al. [30] | Vanilla GAN | |
| | Samangouei et al. [31] | WGAN [45] | |
| | Yu et al. [32] | CGAN [51] | |
| | Zhao et al. [33] | WGAN | |
| Safe image steganography | Hayes et al. [36] | Vanilla GAN | Mainly focus on image processing but lack in security concerned |
| | Zhang et al. [34] | Vanila GAN | |
| | Shi et al. [37] | WGAN | |
| | Volkhonskiy et al. [38] | DCGAN [39] | |
| | Tang et al. [40] | Vanilla GAN | |
| | Zhang et al. [41] | AC-GAN [42] | |
| | Liu et al. [43] | GAN/DCGAN | |
| Neural cryptography | Wu et al. [44] | WGAN | Focus on machine learning and cryptography but on small dataset |
| | Abadi et al. [46] | Vanilla GAN | |
| Security analysis | Chhetri et al. [48] | CGAN | It's only analysis work; does not focus on any specific topic |

**TABLE 5.2**

Security in GANs Network

| References | GAN Type Used in Reference Paper | Purpose of Work |
|---|---|---|
| Gomez et al. [54] | CycleGAN [55] | Cipher Outrageous |
| Hitaj et al. [58] | IWGAN [72] | Password guessing |
| Nam et al. [60] | IWGAN/ RNN | |
| Hu et al. [59] | Vanilla GAN | Malware production and attacks against interruption detection structures |
| Kawai et al. [62] | Vanilla GAN [61] | |
| Rigaki and Garcia [64] | Vanilla GAN | |
| Lin et al. [63] | WGAN | |
| Lin et al. [65] | WGAN | |
| Singh et al. [68] | | |

## 5.4 Security Attacks in GANs

A pre-existing dataset with the discriminator tasked with differentiating between legitimate and malicious communication [52]. This assault is routed via an IDS, and the IDS output is routed through the discriminator, simulating the IDS [53] shown in Table 5.2. Lin et al. obtained favorable consequences, lowering malware recognition rates from 70% to less than 1% [56,57].

Lin et al. [65] used a similar strategy to mitigate denial of service (DoS) charge [66]. The writer altered just those parts of the assault irrelevant to the DoS occurrence and left the efficient components intact. This enabled the writers to cut accurate favorable rates in half to 47.6 percent [67]. A scarcity of properly categorized datasets has hampered the development of malware.

Singh et al. [68] proposed utilizing ACGANs to supplement current malware pictures with additional tagged datasets to solve this problem. AC-GANs improved the stability and robustness of training. With a growing number of samples to aid in training, the augmentation method has improved malware databases.

While many more aspects of cyber security are susceptible to adversarial GAN attacks, these three may offer significant insight into the nature of adversarial GAN charge, and the concepts provided can be extended to other disciplines.

### 5.4.1 Cracking Passphrases

The state-of-the-art password cracking methods now available include calculating millions of hashes from a huge word list and comparing them to the password hashes we are attempting to break. These word collections are often comprised of frequently used or previously used passwords but are not exhaustive. Your password cracking ability is limited by the size of your word list when using this method.

When combined with HashCat, PassGAN could guess between 51 and 73 percent more exclusive passwords than HashCat alone. If these figures are not frightening enough, it is worth noting that PassGAN can generate an almost infinite amount of password guesses. The number of unique passwords created using password generation rules is determined by the sum of regulations and the extent of the password record utilized; however, PassGAN's production is not limited to a tiny portion of the password universe. Consequently, PassGAN could presumption more passwords than the other apparatuses, although they were all skilled on the identical password record.

### 5.4.2 Hiding Malware

Not only is a GAN useful for data generation in cyber security, but it is also capable of avoiding detection systems. The article "Generating Argumentative Malware Illustrations for Black-Box Occurrences Using GAN" delves into this subject in detail. This may be used to create malware that is resistant to detection by machine learning-based detection systems. The concept is similar to a conventional GAN, except that the Black Box detector acts as a discriminator. The generator is fed a mixture of noise and malware samples, and the discriminator is then supplied with benign instances to assist the generator in determining what is "not malware."

### 5.4.3 Forging Facial Detection

I will cover one of the most frequent uses of GANs in picture creation and modification in the last part of the GAN applications section. To be more precise, a GAN may be used to deceive current image recognition systems and generate high-resolution false pictures.

While this chapter is not about cyber security, it does have ramifications for the industry. Utilizing this hostile technology may create fabricated pictures and films of public personalities, other people of interest, and even you. If photorealistic false photographs and movies are quickly produced, this presents a danger to national and personal security. Seeing is no longer believed in this day and age, and this necessitates the rapid development of detection systems capable of distinguishing between genuine and fraudulent.

### 5.4.4 Detection and Response

Detection and reaction are the last topic covered in this chapter. I wish I could tell you that there is a simple technique or a fast cure or software that you can run, but this is not the case. As shown in this article, a GAN's methods are complex and capable of tricking sophisticated systems. So, what are our options? Education and awareness are required to resolve this issue. While any method for detecting and responding to an adversarial GAN is context-dependent, it is essential to understand the GAN's inner workings and be prepared for the worst. GANs will gain prominence soon among individuals working in the area of machine learning-based cyber security. To identify and react to GAN charge, it is essential to design your systems with GANs in mind and not assume machine learning detection is entirely secure. DeepFD is one of the responses created to GAN-produced pictures. The DeepFD researchers developed the system to identify adversarial GAN-generated images that may jeopardize a person's reputation or personal safety [11]. The DeepFD method detects false pictures produced by state-of-the-art GAN networks with a 94.7 percent detection rate.

## 5.5 Conclusion

The survey comprises new GANs research on various topics extending from image steganography and neural cryptography to malware generation, all to train the system to preserve itself more effectively against adverse attack situations, demonstrating the various research occasions for merging NNs and cybersecurity. Additionally, the chapter covers many distinct types of GANs and GAN variants that researchers have utilized to solve essential security situations. It discusses how GANs have improved surveillance in security procedures and system establishment to combat information sensitivity and develop a more secure intrusion recognition system, secure picture steganography, neuronal cryptography, and security examination. Additionally, GANs are utilized to enhance malware and interruption attack performance. The efficacy of GANs is shown by their ability to generate novel and undiscovered charge that may expose defensive system weaknesses. Researchers have taught computer systems to create charge that are difficult to detect and readily evade detection measures. These techniques may then be used to fortify a piece of software against further charge. We have studied the architectures of these systems and the GAN variants that have been utilized, and we have discussed the outcomes obtained in both defensive and attack scenarios. Indeed, GANs should be created for use in evaluating the security strength of goods provided by businesses. GANs may generate tests for a broad range of known and undiscovered threats, resulting in substantially better computers, computer-based goods, and Internet of Things devices. At the moment, the use of GANs insecurity is in its infancy, but it is unlikely to stay that way for long.

## References

[1] "Number of connected devices worldwide 2030—Statista." Available: www.statista.com/statistics/802690/worldwideconnected-devices-by-access-technology/, Aug 22, 2022.

[2] I. K. Dutta, B. Ghosh, and M. Bayoumi, "Lightweight cryptography for the internet of insecure things: A survey," in 2019 IEEE 9th Annual Computing and Communication Workshop and Conference (CCWC), 2019, pp. 475–0481.
[3] K. Khalil, K. Elgazzar, A. Ahmed, and M. Bayoumi, "A security approach for CoAP-based Internet of Things resource discovery," IEEE 6th World Forum on Internet of Things (WF-IoT), 2020, pp. 1–6.
[4] "What is a cyber attack? Recent examples show disturbing trends." CSO. Available: www.csoonline.com/article/3237324/what-is-a-cyberattack-recent-examples-show-disturbing-trends.html, Feb 27, 2020.
[5] "A ransomware attack took The Weather Channel off the air—The Verge." Available: www.theverge.com/2019/4/19/18507869/weather-channelransomware-attack-tv-program-cable-off-the-air, Apr 19, 2019.
[6] "2019 Capital One Cyber Incident — What Happened—Capital One." Available: www.capitalone.com/facts2019/, 2019.
[7] "Ransomware Attack Hits 22 Texas Towns, Authorities Say." *The New York Times*. Available: www.nytimes.com/2019/08/20/us/texas-ransomware.html, 2019.
[8] I. J. Goodfellow, J. Pouget-Abadie, M. Mirza, B. Xu, D. Warde Farley, S. Ozair, A. Courville, and Y. Bengio, "Generative Adversarial Networks," NIPS'14: Proceedings of the 27th International Conference on Neural Information Processing Systems, Volume 2, no. December 2014, pp. 2672–2680, 2014.
[9] B. Ghosh, I. K. Dutta, M. Totaro, and M. Bayoumi, "A survey on the progression and performance of generative adversarial networks," in 2020 11th International Conference on Computing, Communication and Networking Technologies (ICCCNT). Kharagpur: IEEE, July 2020, pp. 1–8.
[10] Z. Pan, W. Yu, X. Yi, A. Khan, F. Yuan, and Y. Zheng, "Recent progress on generative adversarial networks (GANs): A survey," IEEE Access, vol. 7, no. c, pp. 36,322–36,333, 2019.
[11] U. Bergmann, N. Jetchev, and R. Vollgraf, Learning texture manifolds with the periodic spatial GAN, arXiv preprint arXiv:1705.06566, 2017.
[12] N. Jetchev, U. Bergmann, and R. Vollgraf, "Texture synthesis with spatial generative adversarial networks," arXiv preprint arXiv:1611.08207, 2016.
[13] D. Mahapatra, B. Bozorgtabar, and R. Garnavi, "Image super-resolution using progressive generative adversarial networks for medical image analysis," Computerized Medical Imaging and Graphics, vol. 71, pp. 30–39, 2019.
[14] M. Zareapoor, M. E. Celebi, and J. Yang, "Diverse adversarial network for image super-resolution," Signal Processing: Image Communication, vol. 74, pp. 191–200, 2019.
[15] X. Wang, K. Yu, S. Wu, J. Gu, Y. Liu, C. Dong, Y. Qiao, and C. C. Loy, "Esrgan: Enhanced super-resolution generative adversarial networks," European Conference on Computer Vision, 2018, pp. 63–79.
[16] J. Guan, C. Pan, S. Li, and D. Yu, "Srdgan: learning the noise prior for super-resolution with dual generative adversarial networks," arXiv preprint arXiv:1903.11821, 2019.
[17] J. Y. Zhu, T. Park, P. Isola, and A. A. Efros, "Unpaired image-to-image translation using cycle-consistent adversarial networks," Proceedings of the IEEE International Conference on Computer Vision, vol. 2017 October, 2017, pp. 2242–2251.
[18] P. Isola, J. Y. Zhu, T. Zhou, and A. A. Efros, "Image-to-image translation with conditional adversarial networks," Proceedings—30th IEEE Conference on Computer Vision and Pattern Recognition, CVPR 2017, 2017, vol. 2017 January, pp. 5967–5976.
[19] A. Nguyen, J. Clune, Y. Bengio, A. Dosovitskiy, and J. Yosinski, "Plug and play generative networks: Conditional iterative generation of images in latent space," Proceedings—30th IEEE Conference on Computer Vision and Pattern Recognition, CVPR 2017, 2017, vol. 2017 January, no. 1, pp. 3510–3520.
[20] C. H. Lin, C.-C. Chang, Y.-S. Chen, D.-C. Juan, W. Wei, and H.-T. Chen, "Coco-gan: generation by parts via conditional coordinating," in Proceedings of the IEEE International Conference on Computer Vision, 2019, pp. 4512–4521.

[21] "By 2030, each person will own 15 connected devices. here's what that means for your business and content." MarTech Advisor. Available: www.martechadvisor.com/articles/iot/by2030-each-person-will-own-15-connected-devices-here's-what-thatmeans-for-your-business-and-content/, 2018.
[22] McAfee, "McAfee Labs Threats Report," McAfee Labs, Tech. Rep., 2019.
[23] "What to expect from AI and cyber security roles in the future." CCSI. Available: www.ccsinet.com/blog/what-to expect-from-ai-and-cyber-security-roles-in-the-future/, 2016.
[24] W. A. I. R. Development, "Obscuring and Analyzing Sensitive Information with Generative Adversarial Networks," World Wide Technology, Tech. Rep., 2019.
[25] V. Mirjalili, S. Raschka, A. Namboodiri, and A. Ross, "Semi-Adversarial Networks: Convolutional Autoencoders for Imparting Privacy to Face Images," in 11th IAPR International Conference on Biometrics (ICB 2018). Gold Coast, Australia, 2018.
[26] "Perpetual Line Up—Unregulated Police Face Recognition in America." Available: www.perpetuallineup.org/, 2016.
[27] D. P. A. R. Vinchurkar, "A review of intrusion detection system using neural network and machine learning technique," International Journal of Engineering Science and Innovative Technology (IJESIT) Volume 1, Issue 2, November, 2012 , pp. 55–63.
[28] H. Chen and L. Jiang, "Efficient GAN-based method for cyber-intrusion detection," ArXiv, 2019, pp. 1–6.
[29] V Mirjalili et al. "Semi-adversarial networks: Convolutional autoencoders for imparting privacy to face images." 2018 International Conference on Biometrics (ICB). IEEE, 2018. Link: https://arxiv.org/pdf/1712.00321.pdf.
[30] Jun-Yan Zhu et al. "Unpaired image-to-image translation using cycle-consistent adversarial networks." arXiv preprint, 2017. Link: https://arxiv.org/pdf/1703.10593.pdf.
[31] Chih-Chung Hsu and Chia-Yen & Zhuang Lee, Yi-Xiu. Learning to Detect Fake Face Images in the Wild, 2018. Link: https://arxiv.org/ftp/arxiv/papers/1809/1809.08754.pdf.
[32] F. Yu, L. Wang, X. Fang, and Y. Zhang, "The defense of adversarial example with conditional generative adversarial networks," Security and Communication Networks, vol. 2020, pp. 1–12, Aug 2020.
[33] Z. Zhao, D. Dua, and S. Singh, "Generating natural adversarial examples," arXiv preprint arXiv:1710.11342, 2017.
[34] R. Zhang, S. Dong, and J. Liu, "Invisible steganography via generative adversarial networks," Multimedia Tools and Applications, 2019, vol. 78, pp. 8559–8575.
[35] S. Bhallamudi, "Image steganography, final project." Report, Wright State University, Tech. Rep. March 2015.
[36] J. Hayes and G. Danezis, "Generating steganographic images via adversarial training," in Advances in Neural Information Processing Systems, 2017, pp. 1954–1963.
[37] H. Shi, J. Dong, W. Wang, Y. Qian, and X. Zhang, "SSGAN: Secure steganography based on generative adversarial networks," Lecture Notes in Computer Science (including subseries Lecture Notes in Artificial Intelligence and Lecture Notes in Bioinformatics) , 2018, vol. 10735 LNCS, pp. 534–544.
[38] D. Volkhonskiy, I. Nazarov, B. Borisenko, and E. Burnaev, "Steganographic generative adversarial networks," arXiv preprint arXiv:1703.05502, 2017.
[39] A. Radford, L. Metz, and S. Chintala, "Unsupervised representation learning with deep convolutional generative adversarial networks," arXiv preprint arXiv:1511.06434, 2015.
[40] W. Tang, S. Tan, B. Li, and J. Huang, "Automatic steganographic distortion learning using a generative adversarial network," IEEE Signal Processing Letters, October 2017, vol. 24, no. 10, pp. 1547–1551.
[41] Z. Zhang, G. Fu, J. Liu, and W. Fu, "Generative information hiding method based on adversarial networks," in Advances in Intelligent Systems and Computing, August 2020, vol. 905. Springer Verlag, pp. 261–270.
[42] A. Odena, C. Olah, and J. Shlens, "Conditional image synthesis with auxiliary classifier GANs," in International Conference on Machine Learning, 2017, pp. 2642–2651.

[43] J. Liu, Y. Ke, Y. Lei, J. Li, Y. Wang, Y. Han, M. Zhang, and X. Yang, "The reincarnation of grille cipher: A generative approach," arXiv preprint arXiv:1804.06514, 2018.
[44] C. Wu, B. Ju, Y. Wu, N. N. Xiong, and S. Zhang, "WGAN-E: A generative adversarial networks for facial feature security," Electronics (Switzerland), 2020, vol. 9, no. 3, pp. 1–20.
[45] M. Arjovsky, S. Chintala, and L. Bottou, "Wasserstein generative adversarial networks," in 34th International Conference on Machine Learning, ICML 2017, 2017.
[46] M. Abadi and D. G. Andersen, "Learning to protect communications with adversarial neural cryptography," arXiv preprint arXiv:1610.06918, 2016.
[47] L. Monostori, "Cyber-physical systems: Theory and application," in CIRP Encyclopedia of Production Engineering. Springer Berlin Heidelberg, 2018, pp. 1–8.
[48] S. R. Chhetri, A. B. Lopez, J. Wan, and M. A. Al Faruque, "GAN-Sec: Generative adversarial network modeling for the security analysis of cyber-physical production systems," in Proceedings of the 2019 Design, Automation and Test in Europe Conference and Exhibition, DATE 2019. Institute of Electrical and Electronics Engineers Inc., May 2019, pp. 770–775.
[49] P. Jorgensen, *Applied Cryptography: Protocols, Algorithm, and Source Code in C*, vol. 13, no. 3, 2nd ed. New York: John Wiley and Sons Inc., 1996,.
[50] J. Donahue, P. Krahenb ¨ Uhl, and T. Darrell, "Adversarial feature learning," arXiv preprint arXiv:1605.09782, 2016.
[51] M. Mirza and S. Osindero, "Conditional generative adversarial nets," arXiv preprint arXiv:1411.1784, 2014.
[52] I. Kalyan Dutta, B. Ghosh, A. H. Carlson, and M. Bayoumi, "Lightweight polymorphic encryption for the data associated with constrained internet of things devices," in 2020 IEEE 6th World Forum on Internet of Things (WF-IoT), 2020, pp. 1–6.
[53] W. Kinzel and I. Kanter, "Neural cryptography," in Proceedings of the 9th International Conference on Neural Information Processing, 2002. ICONIP'02., vol. 3. IEEE, pp. 1351–1354.
[54] A. N. Gomez, S. Huang, I. Zhang, B. M. Li, M. Osama, and L. Kaiser, "Unsupervised cipher cracking using discrete GANs," arXiv preprint arXiv:1801.04883, 2018.
[55] J.-Y. Zhu, T. Park, P. Isola, and A. A. Efros, "Unpaired image-to-image translation using cycle-consistent adversarial networks," in Proceedings of the IEEE International Conference on Computer Vision, 2017, pp. 2223–2232.
[56] "HashCat—Advanced password recovery." Available: https://hashcat.net/hashcat/, 2020.
[57] "John the Ripper password cracker." Available: www.openwall.com/john/, 2022.
[58] B. Hitaj, P. Gasti, G. Ateniese, and F. Perez-Cruz, "Passgan: A deep learning approach for password guessing," in International Conference on Applied Cryptography and Network Security. Springer, 2019, pp. 217–237.
[59] W. Hu and Y. Tan, "Generating adversarial malware examples for BlackBox attacks based on gan," arXiv preprint arXiv:1702.05983, 2017.
[60] S. Nam, S. Jeon, H. Kim, and J. Moon, "Recurrent GANs password cracker for IoT password security enhancement," Sensors (Switzerland), 2020, vol. 20, no. 11, pp. 1–19.
[61] S. Madani, M. R. Madani, I. K. Dutta, Y. Joshi, and M. Bayoumi, "A hardware obfuscation technique for manufacturing a secure 3d ic," in 2018 IEEE 61st International Midwest Symposium on Circuits and Systems (MWSCAS), 2018, pp. 318–323.
[62] M. Kawai, K. Ota, and M. Dong, "Improved MalGAN: Avoiding malware detector by leaning cleanware features," in 1st International Conference on Artificial Intelligence in Information and Communication, ICAIIC 2019. Institute of Electrical and Electronics Engineers Inc., mar 2019, pp. 40–45.
[63] Z. Lin, Y. Shi, and Z. Xue, "Idsgan: Generative adversarial networks for attack generation against intrusion detection," arXiv preprint arXiv:1809.02077, 2018.
[64] M. Rigaki and S. Garcia, "Bringing a gan to a knife-fight: Adapting malware communication to avoid detection," in 2018 IEEE Security and Privacy Workshops (SPW). IEEE, 2018, pp. 70–75.

[65] Q. Yan, M. Wang, W. Huang, X. Luo, and F. R. Yu, "Automatically synthesizing DoS attack traces using generative adversarial networks," International Journal of Machine Learning and Cybernetics, December 2019, vol. 10, no. 12, pp. 3387–3396.
[66] "Understanding Denial-of-Service Attacks." CISA. Available: https://us-cert.cisa.gov/ncas/tips/ST04-015, Nov 19, 2020.
[67] N. Martins, J. M. Cruz, T. Cruz, and P. H. Abreu, "Adversarial machine learning applied to intrusion and malware scenarios: a systematic review," IEEE Access, 2020, vol. 8, pp. 35 403–35 419.
[68] A. Singh, D. Dutta, and A. Saha, "MIGAN: Malware image synthesis using GANs," Proceedings of the AAAI Conference on Artificial Intelligence, 2019, vol. 33, pp. 10033–10034.

# 6

# *Generative Adversarial Networks-aided Intrusion Detection System*

**V. Kumar**

## CONTENTS



## 6.1 Introduction

The generative adversarial networks (GANs) are deep generative neural network models which are capable of generating data points thatare identical to the data samples of an original data set. GAN uses a generator and a discriminator which are adversarially trained by indulging in a min-max game where the generator's job is to create data samples as close as possible to the data samples of an original data set, and the role of discriminator is to efficiently discriminate real data samples from the fake samples (generated one). The training stops when the discriminator and generator entera state of equilibrium where the generator generates data samples, which are close to the real data samples and the discriminator is confused in distinguishing real and fake data samples. The trained generator can then be used to synthesize data samples, which mimic the data distribution of the original input dataset.

Researchers have employed deep learning techniques to design efficient network intrusion detection systems (IDSs), which are based on anomaly-based detection method where the basic principle of the deep learning model is to detect anomaly in the distribution of the network flows. In order to evaluate their proposed IDSs, researchers use public datasets (CICIDS2017, NSL-KDD, UNSW-NB15 etc.), which contain captured instances of benign and malicious traffic flows and evaluate the performance of their IDS model on those datasets on several parameters like accuracy, false-positive rate, recall, precision and $F_1$ score. However, as the occurrence of intrusion is a low-frequency activity, these datasets suffer from a high imbalance where the samples of benign nature are more in number in

DOI: 10.1201/9781003203964-6

comparison to the number of attack/intrusion data samples. This imbalance challenges the learning process of a deep learning model because the model develops a bias toward normal data samples (as they are more in number). In order to overcome this problem, several statistical oversampling/undersampling techniques have been developed over the years to neutralize the effect of imbalance. To make the number of minority class samples same as the number of occurrences of the majority class sample, these oversampling/undersampling approaches either reducethe number of majority class samples or add afew more numbers of minority class samples. Few of the conventional techniques are synthetic minority oversampling technique (SMOTE), ADASYN, random oversampling, BorderLine smote, KMEans Smote, SVM Smote, etc. The generative capacity of GANs is being put to use by researchers in recent years to address the issue of high imbalance in public datasets. In direct application, the generator of a GAN is used to generate more number of samples of minority class.These generated samples, augmented with original imbalanced dataset, balance the number of malicious and benign data instances. However, the researchers have proposed that augmenting the generated minority class samples directly into an imbalanced dataset might not be efficient as during generation, the original distribution of samples of minority class is not considered leading to the addition of the data samples which might not be valid. This is where the GAN model has an upper hand on the conventional data oversampling techniques thatoverlook the consideration of data distribution and overlap between majority and minority class samples. Researchers have proposed several variations of GANs, which can learn the distribution of normal or benign data samples and effectively generate data samples thatare valid.

GANs are also finding application in designing of intrusion detection models where the generator's reconstruction loss and the discriminator's discrimination loss arebeing used to calculate an anomaly score, which is then used in identifying a test data sample as normal or malicious. Conventional GAN model is not capable of learning the latent space representation of an input dataset and therefore the researchers have used autoencoders and GANs in conjunction to learn the latent space. To use a GAN as a classifier, an autoencoder and a GAN are trained together to exclusively learn a latent space representation of normal/benign data samples. Once the latent space is learned, the autoencoder can be used to reconstruct a test sample, which if an anomaly, will produce a high reconstruction error and can be labeled as an anomaly. To further improve the detection, both the generator's reconstruction loss and the discriminator's discrimination loss are used to arrive at a reasonable anomaly score.

To verify the robustness of a machine learning-based IDS, the IDS is fed with adversarial examples to influence its learning and fooling it to classify attack samples as normal ones. If the IDS model can easily be manipulated using adversarial examples, it can be considered as a weak IDS. Researchers are using GANs to produce adversarial examples. In the majority of the datasets, the attack samples have functional and non-functional features. The functional features of a sample are attack-specific features, which, if modified, will lead to the generation of invalid data samples; however, the non-functional features can be manipulated in such a way that the attack's integrity is intact but the sample may evade detection. The technique is used by GANs where the generator learns to generate adversarial data samples, which are malicious in nature but have only the non-functional features modified in such a way that an IDS classifies them as benign. As the IDS might not be known to the attacker, a black-box IDS is considered in its place. Based on IDS classification findings on various normal and benign data samples, the discriminator is utilized to learn and emulate the black-box IDS. The discriminator indicates the generator whether the black-box IDS correctly classifies generated samples by the generator; if yes, the generator improves itself to produce more stealthy attack samples, which might evade detection by the IDS. When the generator and discriminator reachNash equilibrium, the generator is able to generate attack samples which are able to evade detection by the IDS.

The rest of the chapter is organized as follows: the application of GANs for data imbalance is discussed in Section 6.2 followed by a review of research work conducted in application of GANs as classifier in Section 6.3. Section 6.4 discusses the use of GANs for designing adversarial networks, which generate adversarial examples with an aim for misclassification by an established IDS. Finally, a brief conclusion of the discussed works is presented in the Conclusion section.

## 6.2 Application of GANs for Resolving Data Imbalance

Over the years, several researchers have proposed data augmentation technique using GAN to resolve data imbalance. In simple terms, the generator of a trained GAN is used for generating minority class samples, which are augmented with original imbalanced dataset to get a hybrid data set containing both the original samples and the generated ones. The hybrid dataset now contains the same number of benign and malicious data samples. A machine learning classifier is then trained on the hybrid dataset, which learns the parameters without any bias towards benign data samples yielding an effective detection model. Results have shown excellent improvement on the accuracy of the classification model trained on the hybrid data set.

In [1], the authors have proposed the use of GANs for addressing the problem of dataimbalance in CICIDS2017 dataset. The rare class data is fed to GAN model to generatesufficient number of rare class samples, the generated data in addition to the data other than rareclass makes the training dataset. A random forest classifier is then trained using the trainingdataset and effective results were obtained by the authors. The authors have also comparedthe result of their proposed method with SMOTE based data augmentation technique andhave verified the effectiveness of GANs for resolving data imbalance and improving machine learning task. To address the issue of cyber attacks in cyber physical systems (CPS) and Internet of Things (IoT), authors in [2] have proposed an IDS based on deep learningtechniques. However, as mentioned by the authors, the state-of-the-art CPS technologies are inflicted with imbalance and unavailable data samples, which challenge the training of the IDS.

To overcome the issue of data imbalance, the authors have used an interesting scheme of data augmentation where a GAN model is used for generating data samples of minority class but only those generated samples which improve the accuracy of the IDS model are incorporated with the samples of the original dataset. As illustrated in Figure 6.1 the authors have employed four modules in the proposed IDS:a data collection module, a data synthesizer module, an IDS module and a controller module. The data collection module collects the real-time data from a CPS network traffic and stores it in a local database; the database contains both the synthetic (generated) data samples and the actual collected data samples. In the event of data imbalance, the data collection module communicates with the data synthesizer module to synthesize data samples belonging to the minority class. However, these generated samples are not readily added into the database; the controller module first verifies whether the addition of the generated sample improves the overall accuracy of the IDS model. Only when the addition of the synthetic sample improves the accuracy

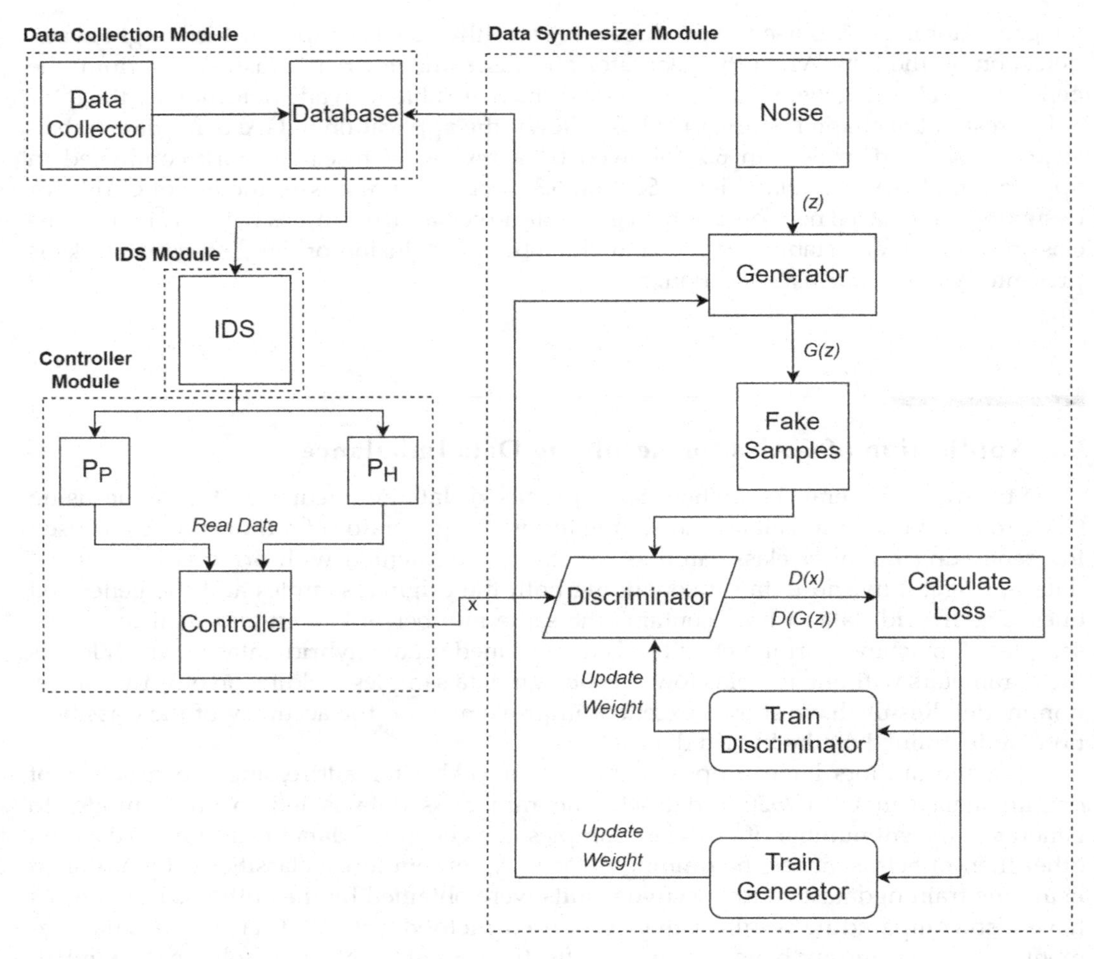


**FIGURE 6.1**
Framework of G-IDS.

of the IDS, it is actually added to the database of the data collector module; otherwise, it is rejected. The authors have tested the execution of the proposed IDS using NSL KDD-99 dataset and have compared the results with non-GAN-assisted IDS. The published results validate the improvement in the performance of the GAN-assisted IDS on various metrics like precision, recall and F-1 score over a simple IDS without GAN assistance.

In [3], the authors have used a multiple fake class generative adversarial network (MFC-GAN) to handle the class imbalance problem. In contrast to the conventional GAN model, which uses a one fake class, the proposed MFC-GAN generates multiple fake class samples. The GAN model is trained to synthesize samples of minority class to rebalance the dataset. The synthetic data samples are augmented with the original dataset and a convolution neural network is trained using the new dataset. The fine-grained training of the discriminator of MFC-GAN enables the discriminator to easily classify synthesized data into multiple fake-classes and minimizes the risk of classifying generated data into actual classes. In other words, the MFC-GAN classifies the real data samples as real class and generated samples as multiple fake-classes, which enhance the performance of the generator to synthesize class-specific data samples. The objective function of the proposed model is to maximize the expectation of classifying real data samples into actual class $C$ and classifying fake data samples into multiple fake classes $C'$as is shown in the following equations:

$$L_S = E[\log P(S = real \mid X_{\text{real}})] + E[\log P(S = fake \mid X_{\text{fake}})] \tag{6.1}$$

$$L_{cd} = E[\log P(C = c \mid X_{\text{real}})] + E[\log P(C' = c' \mid X_{\text{fake}})] \tag{6.2}$$

$$L_{cg} = E[\log P(C = c \mid X_{\text{real}})] + E[\log P(C = c \mid X_{\text{fake}})] \tag{6.3}$$

$L_s$ is the loss of sampling, which is actually the sample's probability to be real or unreal. $L_{cd}$ and $L_{cg}$ are discriminator and generator classification loss. The authors have compared the results obtained using MFC-GAN on MNIST dataset with several other class imbalance resolving methods like SMOTE, AC-GAN, FSC-GAN and MFC-GAN. The MFC-GAN model has performed fairly well in sensitivity, specificity, accuracy, precision, recall, F1-score and G-mean.

In [4], the authors have proposed an IDS, which uses a conditional Wassersteingenerative adversarial network (CW-GAN) and a cost-sensitive stacked autoencoder(CSSAE) to detect known/unknown attacks. CWGAN generates minority class data samples, which are then augmented with the actual dataset to be used for training by aCSSAE. The authors have used a gradient loss and $L_2$ regularization term in CWGAN toresolve the issue of mode collapse and gradient convergence. The loss function of CSSAEgives a higher cost for misclassifying the data points of minority class, which improves therecognition of the datasamples of minority class. The authors have combined a conditional GAN and Wasserstein GAN to enhance the quality of the synthesized samples; also the authors have claimed that use of CGWAN can eradicate the issue of mode collapse and gradient convergence which are to an extent, inherit problems in GANs. Figure 6.2 illustrates the framework of the proposed IDS. The framework is composed of three modules: the CWGAN, data augmentation and attack detection module. CWGAN has a discriminator and generator, which are trained in adversarial fashion with class label $y$, which enables the generator to generate class-specific data samples. The data augmentation module uses the learned generator to synthesize samples of the minority class, which are augmented with attack samples of original data samples to generate a new data for training. The new training data is then used to train a cost-sensitive stacked autoencoder, which classifies the data samples into malicious/attack or normal class.

The conventional stacked autoencoder (SAE) assumes that the misclassification cost of each class is same and therefore performs poorly for classification of data samples of minority class. To overcome this problem, the authors have given a large wrong-classification penalty to the data of minority class and a low wrong-classification penalty to the data of majority class in the original loss function of SAE. This modified loss function is used in the CSSAE, which improves its learning for the classification of the minority class data samples.

The authors have assessed the proposed IDS on benchmark datasets of NSL-KDD and UNSW-NB15 using several performance metrics like detection rate, F1 score, G-mean and false-positive rate. On all of the benchmark datasets, the authors claim that the proposed CSSAE network has a rapid convergence speed, a minimal training error and a high accuracy.

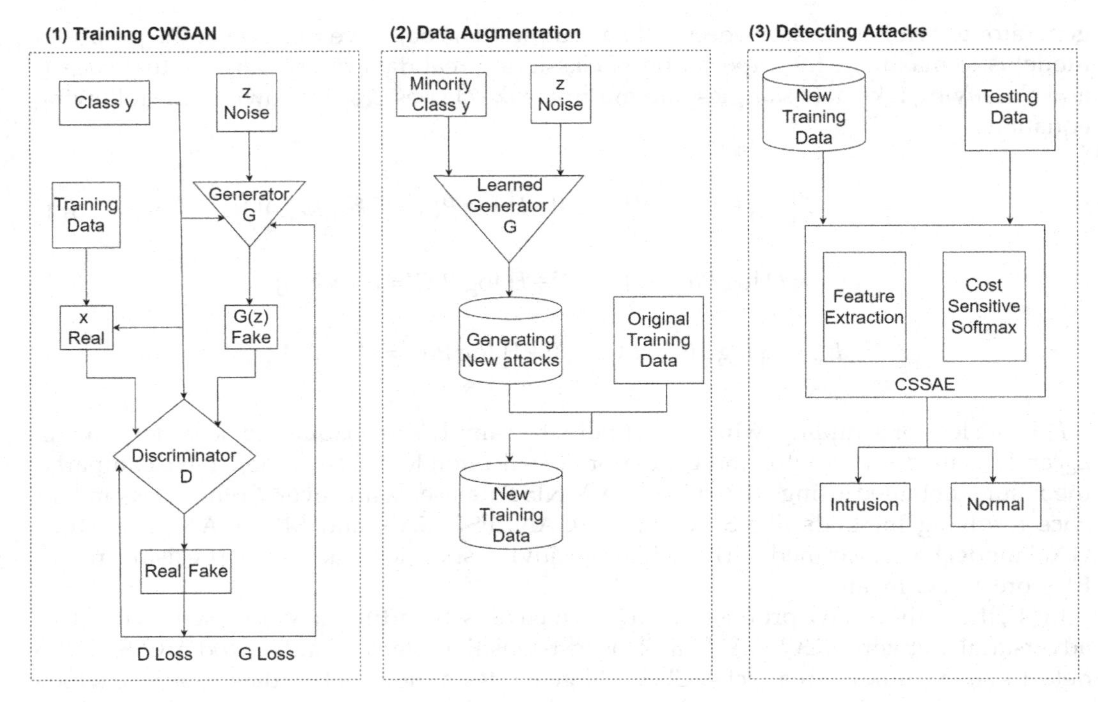


**FIGURE 6.2**
Proposed intrusion detection framework.

The authors have also compared the proposed data augmentation technique against several other methods, which includes random oversampling (ROS), synthetic minority oversampling technique (SMOTE) and ADASYN (adaptive synthetic). The published results validate that the proposed data augmentation technique has outperformed the ROS, SMOTE and ADASYN data techniques on detection accuracy of specific attacks and have also outperformed these methods on accuracy, precision, DR, F1 score, G-mean and false-positive rate for benchmark dataset of NSL-KDD and UNSW-NB15.

In [5], the authors have used a *K*-nearest neighbor (KNN) and GAN-based technique for detecting intrusion in an imbalanced dataset. The majority class data samples are undersampled using KNN algorithm and minority class samples are oversampled using a tabulated auxillary classifier generative adversarial network model (TACGAN) to balance the dataset. Figure 6.3 illustrates the TACGAN-IDS framework, which has three functional modules:data preprocessing, TACGAN and undersampling module. The data preprocessing module pre-processes the input data by performing: (1) encoding of the nominal features and (2) data normalization of all the numeric features to (0,1) interval. Also a deep autoencoder is used post data preprocessing to extract only the important features and remove the redundant features. The authors have efficiently used KNN algorithm for under sampling the majority class samples. Let $X= \{x_1,x_2,x_3,\ldots x_n\}$ be $n$ data points where the label for each data point is known. For a given point $x \in X$, $k$ nearest point to point $x$ are denoted as $X'=\{x_1,x_2,x_3,\ldots x_k\}$. The point $x$ is assigned a class, which occurs the most in $X'$. The authors have considered the dataset which suffers from overlapping where few data points of the majority class and minority class data points share a common subspace. The overlapping data points pose a challenge for the classification algorithm as

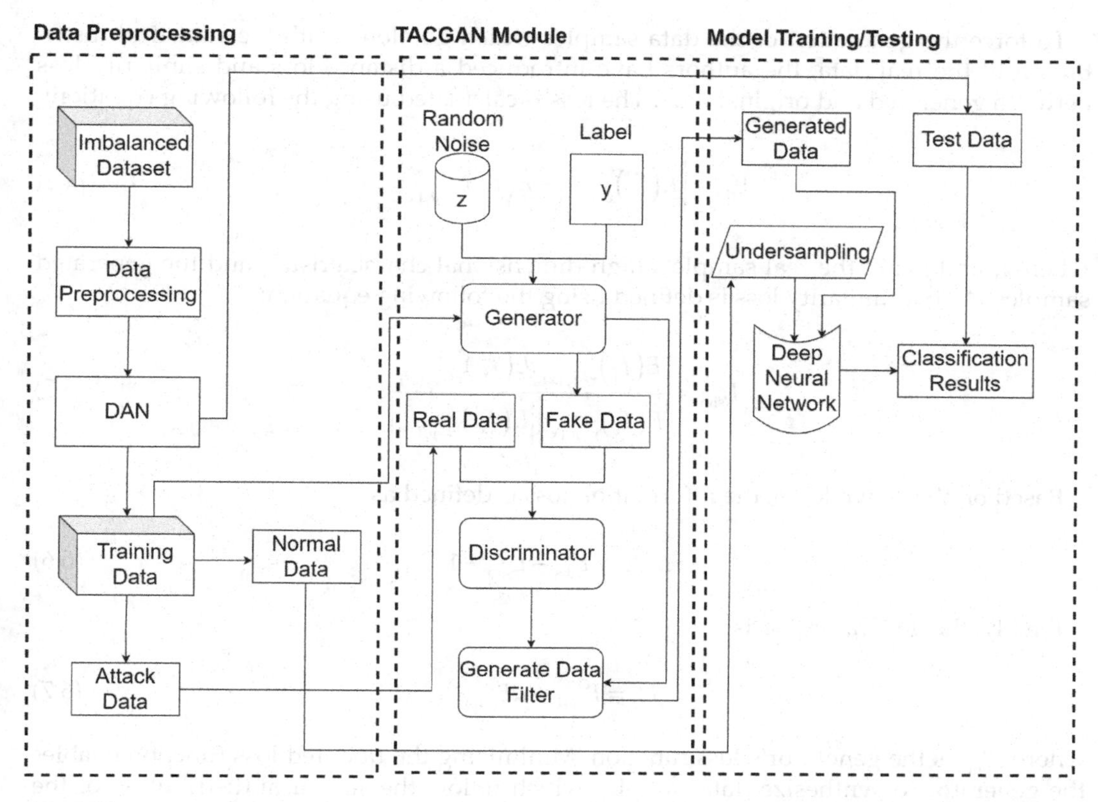


**FIGURE 6.3**
TACGAN-IDS model framework.

it struggles to find a viable decision boundary. To solve this problem, the authors have divided data samples into outliers, boundary data and trusted data. For each test data point,fivedata points with a minimum distance to the test point are chosen. The number of similar class data points in fivesamples is denoted by *s*.

If $s = 0$, it means that the test data point is surrounded by all the data points of different class, the test sample in this case is marked as an outlier. If $s = 1, 2, 3, 4$, it means that the test sample is surrounded by data points of both the classes, the test sample is therefore marked as a boundary data. If $s = 5$, it means that the test data sample is surrounded by the data points of the same class and is therefore marked as trusted data. The marking scheme is performed for all the majority class samples followed by undersampling of outliers and boundary data.

For oversampling of minority class data samples, TACGAN module is used, which generates the minority class data samples. TACGAN is a conditional generative network, which is trained by iteratively training a discriminator and a generator as similar as a GAN with a difference that now the generator and discriminator are also trained on class labels. Once the TACGAN model is trained, the generator is used to synthesize the data points of the desired minority class. Also, the proposed framework has a filter, which uses the discriminator to filter out the generated samples, which are inferior so as to only retain those generated samples that, to some degree, areequivalent to actual data points.

To force the generation of the data samples which go along with the statistical distribution of the real data, the authors have introduced a distance loss and similarity loss between generated and original data. The loss is calculated using the following equation:

$$L_{\text{dis}} = \left\| E(f_x)_{x \sim p_r(x)} - E(f_{x'})_{x' \sim p_g(x')2} \right\| \quad (6.4)$$

where $f_x$ and $f_{x'}$ are the real sample's high dimensional characteristics and the generated samples. Also a similarity loss is defined using the following equation:

$$L_{\text{sim}} = \frac{E(f_x)_{x \sim p_r(x)} \cdot E(f_{x'})_{x' \sim p_r(x')}}{\left|E(f_x)_{x \sim p_r(x)}\right| \left|E(f_{x'})_{x' \sim p_r(x')}\right|} \quad (6.5)$$

Based on these two loses, the information loss is defined as

$$L^G{}_{\text{dev}} = L_{\text{dis}} - L_{\text{sim}} + 1 \quad (6.6)$$

Finally, the generator loss is

$$L_G = L^G{}_{\text{ori}} + \alpha L^G{}_{\text{dev}} \quad (6.7)$$

where $L^G{}_{\text{ori}}$ is the generator's loss function. Minimizing the updated loss function enables the generator to synthesize data samples which follow the statistical distribution of the real dataset.

The authors have evaluated the proposed IDS framework on benchmark datasets including KDDCUP99, UNSW-NB15 and CICIDS2017 using several performance metrics like accuracy, precision, F-measure, recall and false-positive rates. The authors have compared the classification performance of their proposed method against other baseline methods including random oversampling, SMOTE, SMOTE+ENN, ADASYN, MAGNETO, CGAN, ACGAN and WCGAN. For KDDCUP99, the proposed method outperforms all the baseline method in accuracy, recall and F1-score. For UNSNW-NB15, the results are similar except for recall score where it is behind MAGNETO method.

## 6.3 Application of GAN as a Deep Learning Classifier

The previous section discussed the role of GAN in mitigating data imbalance in existing public datasets. However, the role of GAN was limited to being a data generator and to further fulfil the goal of actual classification of network traffic, separate machine learning/deep learning classifiers were used. In other words, the data imbalance was resolved by GAN and classification was carried out by a separate classifier. Over the years, researchers have developed several techniques through which a GAN can itself act as a classifier for intrusion detection.

Anomaly detection issue is discussed by authors in [6] who have proposed an anomaly detection technique for CPSs where a GAN model is trained on a sequence of normal traffic data and the generator learns to generate data samples as similar to the normal data samples. The trained generator and discriminator are then fed with test data samples and the corresponding reconstruction and discrimination losses are calculated. If the calculated loss exceeds a pre-defined threshold, the test sample is detected as an anomaly. The logic stems from the observation that the GAN model is exclusively trained on benign traffic data and therefore has a trained generator and discriminator, which have aligned themselves with the data of benign class. Therefore, if at test time an anomaly data sample is fed to the generator or discriminator, it is highly likely that it will generate a high value of reconstruction error or discrimination value, which can be used as an anomaly score.

The trained generator is efficient in synthesizing real-looking data samples from random noise as it is a feature mapping to real space X from a latent space Z,$G(Z): Z \rightarrow X$, and therefore it gets tricky to compute the reconstruction loss of a test data sample $X^{test}$ as first the latent representation of $X^{test}$ is to be searched in Z. The authors have proposed a heuristic method for searching the latent space depiction of $X^{test}$in Z. To find the latent depiction$Z^k$ of $X^{test}$ a random set $Z^1$ is sampled from the latent space and the corresponding reconstruction error is computed using G($Z^1$) by feeding the $Z^1$to the generator. The sample is updated after obtaining gradients from the following loss function between $X^{test}$ and *G(Z)*.

$$\min_{Z^k} Er\left(X^{\text{test}}, G_{rnn}\left(Z^k\right)\right) = 1 - Simi\left(X^{\text{test}}, G_{rnn}\left(Z^k\right)\right) \tag{6.8}$$

By iteratively updating $Z^k$ using the above loss function, the algorithm inches closer to the most suitable latent representation $Z^k$ of $X^{test}$ . Once $Z^k$has been found the reconstruction residual for testing samples at time *t* can be computed as

$$\text{Res}\left(X_t^{\text{test}}\right) = \sum_{i=1}^{n} \left| x_t^{\text{test},i} - G_{rnn}\left(Z_t^{k,i}\right) \right| \tag{6.9}$$

As the discriminator has been trained to distinguish between real and fake/anomaly samples, it may be used for detecting anomalies. The reconstruction loss of a test sample at time $t, Res\left(X_t^{\text{test}}\right)$, and discrimination loss $D_{rnn}\left(X_t^{\text{test}}\right)$ are then used to calculate a discrimination-reconstruction score called DR-score (DRS). A test sample with high DR-score is labelled as an anomaly. Figure 6.4 illustrates the framework of the anomaly detection that is proposed using GAN. The left portion of the figure depicts the model training where the discriminator and generator are trained on real time CPS data. The right part of the figure depicts the functioning of the model as an anomaly detector where the input test samples undergo invert mapping to retrieve their latent space representation after which the reconstruction loss and discrimination loss arecomputed as described earlier. Finally, the DRS score is computed to predict the label of the test sample as benign or anomaly.

As can be observed, the central idea of anomaly detection using GAN revolves around learning the latent space representation of data samples, which is then used to compute reconstruction losses to arrive at an anomaly score. Most of the research work in the area is dedicated to finding ways of learning the latent space representation of the generated data samples. In [7], the authors have suggested an unsupervised anomaly-based IDS for CPSs, which uses a GAN and an autoencoder to understand the latent space of the data samples. In contrast to the latent space learning technique proposed in [6], authors in [7]

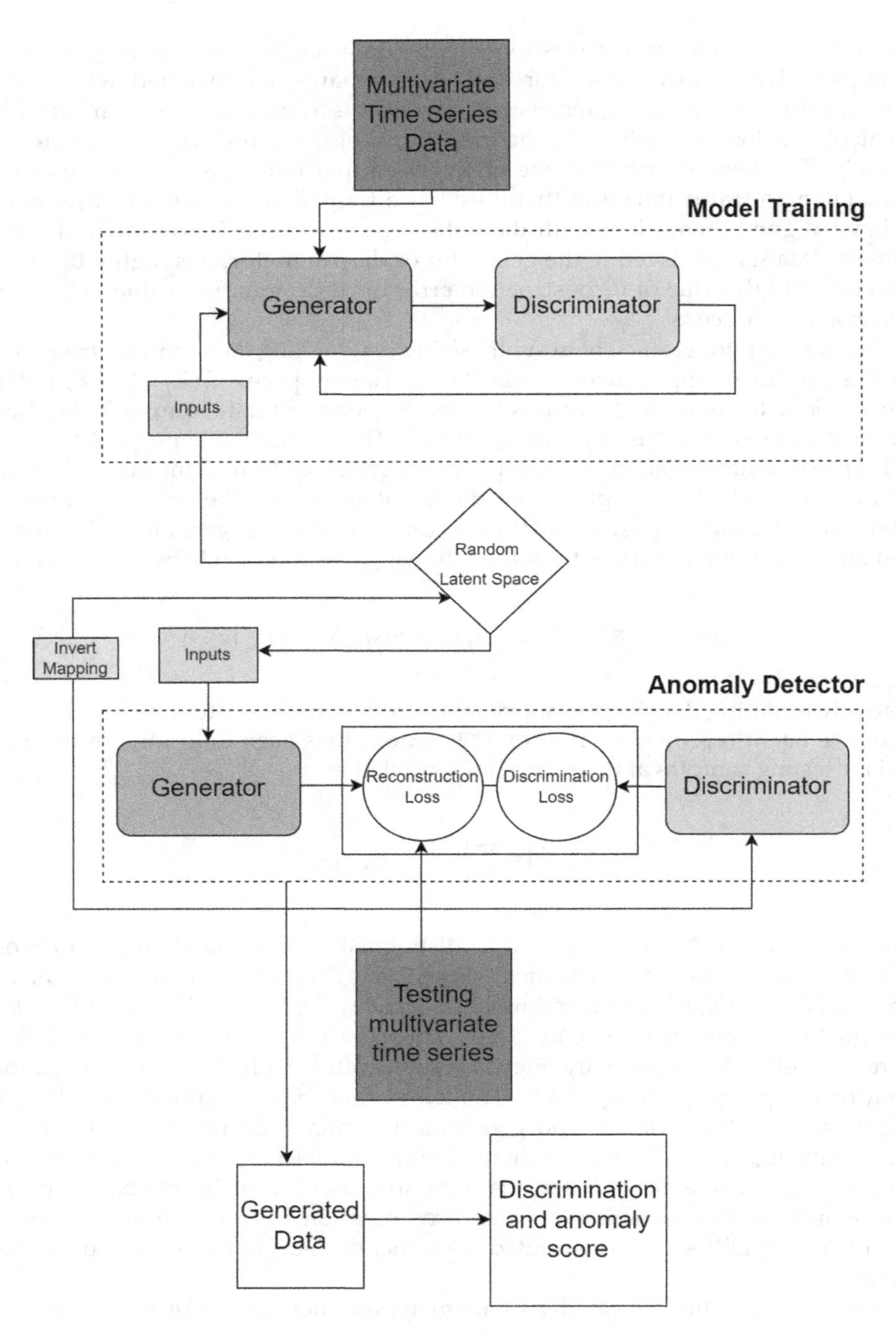


**FIGURE 6.4**
Proposed GAN-based anomaly detection framework.

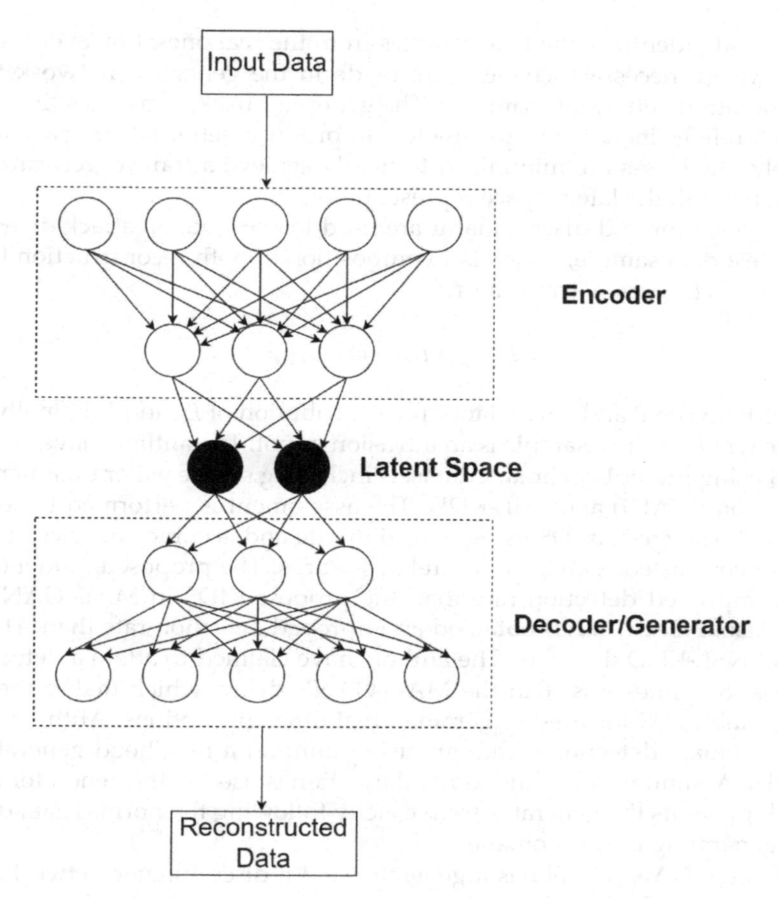


**FIGURE 6.5**
Autoencoder to learn the latent space.

have trained an autoencoder alongside the discriminator and generator where the role of autoencoder is to learn the latent space $Z$.

As discussed earlier, the generator of a GAN gives mapping of noisy data points in $Z$ to the real data space, but it is not capable of doing the reverse, which is the mapping between data space and latent space. This requires the generator function to be invertible, which is generally not the case. Therefore, the authors have proposed the use of an encoder $E$, which is trained to perform the mapping from data space $X$ to $Z$. Thus $E$ is trained to do the mapping:

$$E(x): X \rightarrow Z$$

Figure 6.5 shows the autoencoder in which the encoder part contributes tolearning the latent space. The autoencoder's decoder part is actually the generator of a GAN. The encoder part is inputted with data and it tries to form a constraint depiction of the data (the encoded data) and the decoder/generator regenerate the data from the latent space. So, now in place of generating data samples from random noise, the generator generates data samples from the latent space. Initially, the generated samples are weak and the discriminator easily identifies the fake samples from the real ones; however, as the losses are backpropagated, necessary changes are made in the generator network to make it capable of generating better data samples. The generator backpropagates the error to the encoder, which refines its network parameters to produce better latent space representation. Iteratively, the losses are minimized to finally achieve a trained generator and discriminator, along with the latent space representation $Z$.

The trained generator and discriminator are used to compute an attack detection score ($AD_{\text{score}}$) for a test data sample, which is a composition of both reconstruction loss $L_R$ and discrimination loss $L_D$ parametrized by $\tau$.

$$AD_{\text{score}} = \tau L_D + (1 - \tau) L_R \tag{6.10}$$

where $\tau$ varies between 0 and 1 to balance the contribution of $L_R$ and $L_D$. Finally, $AD_{\text{score}}$ is used to decide whether a test sample is an intrusion or not. The authors have evaluated the proposed IDS using public benchmark datasets including secure water treatment (SWAT), water distribution (WADI) and NSL-KDD. The assessment is performed to compute the competence of the suggested IDS using detection rate and accuracy metrics; the obtained results are also contrasted with previous related works. The proposed model (FID-GAN) has achieved improved detection rate than the proposed IDS in MAD-GAN [6] for all the datasets. Also, FID-GAN has obtained an improved detection rate than ALAD [8] for the WADI and NSL-KDD data sets. The authors have claimed to attain a detection delay, which is atleast 5.4 times less than the MAD-GAN's delay, which makes the proposed FID-GAN a suitable IDS for time constrained real time applications. Authors in [9] have proposed an anomaly detection technique using minimum likelihood generative adversarial networks. A minimum likelihood regularization is used in the generator network of a GAN, which prevents the generator from exactly following the normal data distribution and helps in generating more anomalies.

The conventional GAN model has a generator and a discriminator network, which are adversarially trained and post training, the generator is able to synthesize data samples from a random noise, which mimic the real data samples. In doing so, the discriminator degenerates in the final part of the training where it outputs 0.5 for the synthesized data by the generator. To prevent the degeneration of discriminator $D$ and to enhance the discriminator's performance during training, the authors have proposed to regularize the generator $G$ such that: (1) $G$ generates more anomaly data samples during the training and (2) the random data distribution of the generator, $p_G$, does not coincide to real data distribution $p_{\text{data}}$. For achieving this, the authors have used a KL divergence term for regularizing the objective function of the generator, the divergence term $KL(p_{\text{data}} \,||\, p_g)$ is minimized so as to make $p_g$ have low values on normal data. This type of GANs are known as MinLGAN. The introduction of the KL divergence term facilitates the generator to synthesize more data points around the normal data subspace, which consequently doesn't allow $G$ to coincide to real distribution. The performance of the anomaly detection method (MinLGAN) is evaluated using CIFAR10 dataset. On ROC scores, MinLGAN performs better than GAN on class zero, two, four, five, seven, eight and nine of CIFAR10 dataset. To overcome the restriction of the discriminator $D$ in the conventional GAN to only act as a binary classifier, authors in [10] have proposed to modify the GAN framework to support multiclass classification. Thus, the ability of conventional discriminator to detect the input data sample as real or fake is further extended for also classifying the actual class of the real data samples. The discriminator $D$ of GAN is replaced by multiclass classifier $C$, as it now classifies the input data into one of the sixpossible categories (normal, dos, probe, R2L, U2R and fake) as depicted in Figure 6.6. The training of

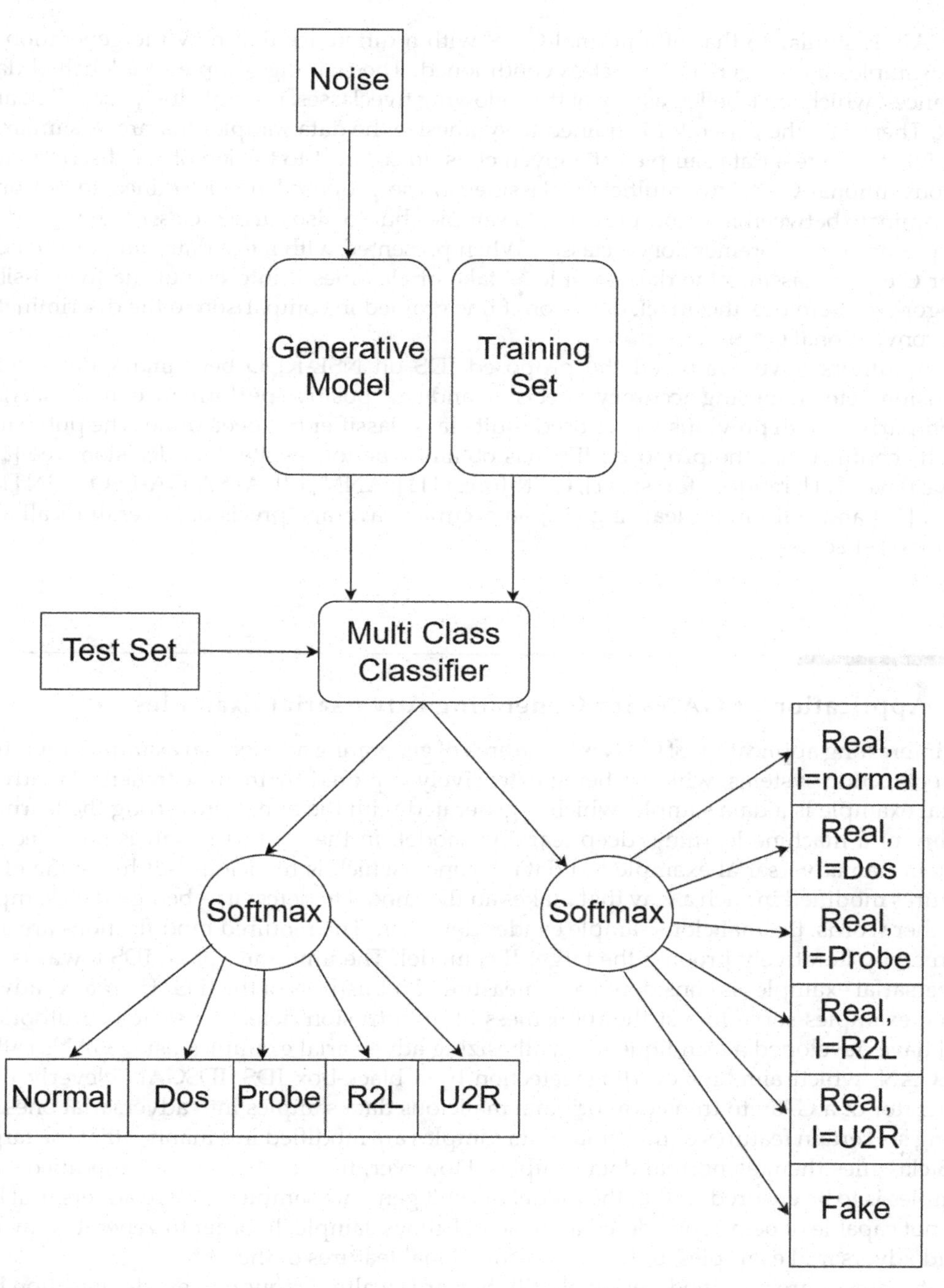


**FIGURE 6.6**
Framework of ID-GAN.

ID-GAN is similar to that of a normal GAN with a difference that now the generation of data samples by the generator is class conditioned. The training samples are labelled data instances, which are labelled as any of the following fiveclasses: normal, dos, probe, R2L and U2R. Therefore, the generator is trained to synthesize the data samples thatare as similar as possible to the real data samples of a given class. In contrast to the job of the discriminator in conventional GAN, the multiclass classifier in the proposed IDS is trained to not only discriminate between real and unreal data samples but to also further classify the real data samples into the aforementioned classes. When presented with a test data sample, the classifier *C* either classifies the data sample as fake or classifies it into one of the fivepossible categories. Therefore, the attack detection if fine grained in comparison to the discriminator of a conventional GAN.

The authors have evaluated the proposed IDS on NSL-KDD benchmark dataset for intrusion detection using accuracy, precision and true positive performance metrics. Also, a comparison with previously proposed multiclass classifier has been done. The published results confirm that the proposed IDS has obtained better results than decision tree [11], naive Bayes [11], random forest [11], GAR-forest [11], ANN [12], AFSA-GA-PSO-DBN [13], CNN [14] and self-taught learning [15] on accuracy, average precision, overall recall and average F-1 score.

## 6.4 Application of GANs for Generating Adversarial Examples

An interesting application of GAN is in the area of generating adversarial examples for intrusion detection systems, which is being extensively explored by the researchers. An adversarial example is a data sample, which is generated with the aim of thwarting the learning ability of a machine learning/deep learning model. In the context of intrusion detection systems, an adversarial example is a data instance, which is malicious but has some of its features modified in such a way that makes an IDS model to detect it as benign data sample; in other words, the malicious sample evades detection. The required modifications are also learned by iteratively probing the target IDS model. The immunity of an IDS towards the adversarial examples is considered as a measure of robustness of the IDS. Therefore, adversarial examples serve to test the robustness of an intrusion detection system. Authors in [16] have developed a technique for synthesizing adversarial examples using GANs called IDSGAN, which aims for evading detection by a black-box IDS. IDSGAN cleverly uses generator of a GAN to transform original malicious data samples into adversarial ones. In doing so, certain features of malicious data samples are modified in a manner that the target IDS classifies them as normal data samples. However, the validity of such modified data samples is to be ensured so that the model doesn't generate samples thatare adversarial but are not capable of being considered as an actual attack sample. In order to generate only the valid adversarial examples, only the non-functional features of the

data sample are modified iteratively till they are stealthy enough to evade detection but are also valid.

As the actual IDS model is generally assumed to be not available, a black-box IDS model is implemented to simulate a real-life IDS. The proposed IDSGAN model is based on Wasserstien GAN where the generator is trained to only modify certain non-functional feature of a data sample to generate adversarial examples. Depending on the nature and purpose of a given attack, the functional features of attack samples may vary. These

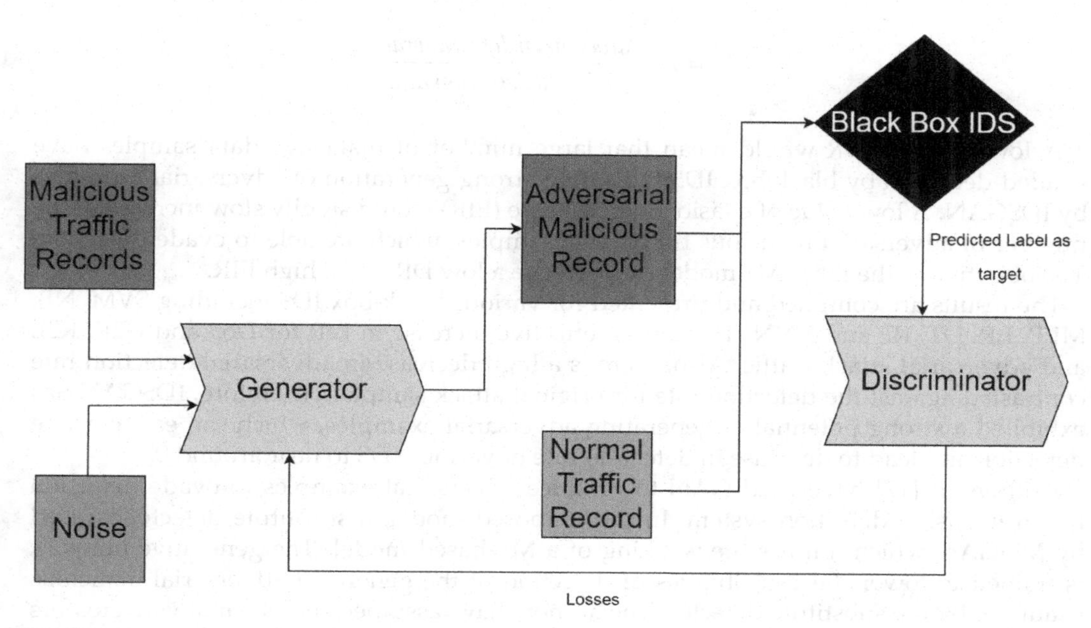


**FIGURE 6.7**
Training of IDSGAN.

functional features are the set of features thatform the "nucleus" of a given attack and are therefore not modified for generating adversarial data samples. The generator is trained to only modify the features of an attack category data sample, which are non-functional. The discriminator's role is to simulate the black-box IDS and provide feedback for the generator training.

The results which the black-box IDS outputs on the identified samples are imitated by the discriminator. The discriminator enhances the generator's training as the loss is computed based on the decision of discriminator and is backpropagated to the generator. The generator therefore attacks the discriminator and tries to fine-tune the network parameters to generate adversarial examples, which can fool black-box IDS into misclassification. Figure 6.7 depicts the training of the IDSGAN model. The generator takes malicious traffic record and random noise to generate adversarial network traffic. The generated adversarial traffic records and normal traffic records are both fed to the IDS and the discriminator. The IDS outputs the predicted label for the input data samples and the discriminator trains itself based on the prediction of the IDS. The discriminator feedbacks the loss gradient to the generator, which then improves itself to generate better adversarial examples. To put it simply, the discriminator indicates the generator whether the black-box IDS has been fooled by the adversarial examples that it generated. If that is not the case, the generator updates the network parameter to produce better adversarial samples. The training continues till maximum possible adversarial examples are predicted as normal traffic by the black-box IDS.

The following performance metrics are used:

$$DR = \frac{No.of\ correctly\ detected\ attacks}{No.of\ all\ attacks}$$

$$EIR = 1 - \frac{Adversarial\,detection\,rate}{Original\,detection\,rate}$$

A low value of $DR$ would mean that large number of malicious data samples have evaded detection by black-box IDS indicating strong generation of adversarial examples by IDSGAN. A low value of evasion increase rate (EIR) would signify slow increase in the number of adversarial malicious traffic data samples, which are able to evade detection. The objective of the IDSGAN model is to achieve a low DR and a high EIR.

The results are compiled and presented for various black-box IDS including SVM, NB, MLP, LR, DT, RF and KNN. There is an effective increase in EIR for DoS and U2R, R2L and adversarial attack traffic. Also, there is a high decrease in adversarial detection rate contrasted against the detection rate for original attack samples. Therefore, IDSGAN has exhibited a strong potential of generating adversarial examples, which can easily evade detection and lead to decrease in detection rate of various IDS to near around 0.

Authors in [17] have used GAN to produce adversarial examples to evade detection by an intrusion detection system. In the proposed model, a substitute detector is used by MalGAn, which mimics the working of a ML-based model. The generative network is trained to lower the probabilities of detection of the generated adversarial malicious examples by the substitute detector. The authors have assumed that the malware creators only know the features that the IDS uses and are unaware about the type of IDS or its internal workings. The model is comprised of a substitute detector and a generator both of which are FFNN. Figure 6.8 shows the architecture of MalGAN. The generator converts a given malware feature vector into an adversarial feature vector. As input, it accepts a malware feature vector $m$ and a noise vector $z$ and concatenates them. Each element in the noise vector $z$ is sampled from a uniform distribution [0,1]. This input vector is then fed to the generator of the network, which outputs the generated adversarial instance of the input vector as $o'$. The discriminator of the GAN is used as the substitute detector, whose purpose is to mimic the black-box detector. The substitute detector takes the feature vector $x$ and outputs whether the input instance is a malicious program or a benign one. The training data, which is fed to the substitute detector, contains both the adversarial malicious examples generated by the generator and a bunch of benign data samples, which are carefully selected by the malware creators.

The substitute detector's loss function is defined as

$$L_D = -E_{x \in BB_{\text{Benign}}} \log\left(1 - D_{\theta_d}(x)\right) - E_{x \in BB_{\text{Malware}}} \log\left(D_{\theta_d}(x)\right) \tag{6.11}$$

where $D_{\theta_d}(x)$ is probability of the substitute detector to classify $x$ as a malware, $BB_{\text{Benign}}$ is the set of data instances labelled by black-box detector as benign and $BB_{\text{Malware}}$ is the data instances which the black-box detector labels as malware. For training, the loss $L_D$ should be minimized with the weight of the feed-forward neural network. Similarly, the generator's loss function is defined as

$$L_G = E_{m \in S_{\text{MALWARE}}, z \sim P_{uniform[0,1)}} \log D_{\theta_d}(G_{\theta_g}(m, z)) \tag{6.12}$$

where $S_{\text{MALWARE}}$ is the actual dataset containing the malware instances. $L_G$ is minimized against the weight of the feed-forward neural network of the generator. The objective of minimizing $L_G$ is to ensure that the probability of the substitute detector to detect a

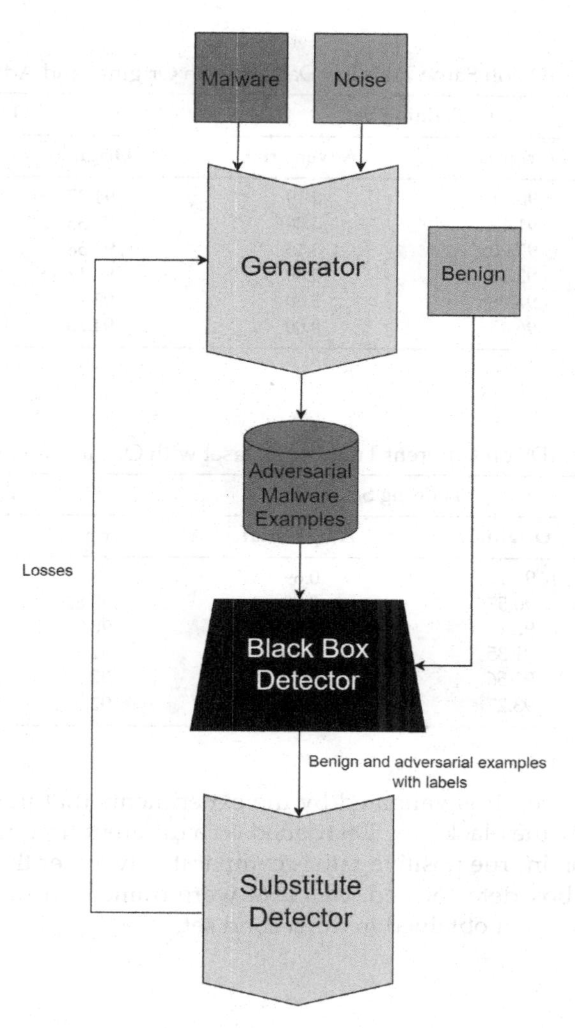


**FIGURE 6.8**
The architecture of MalGAN.

malware sample is reduced iteratively to a point where it predicts the malware sample as benign. As the substitute detector is mimicking the actual detector, the subsequent training will enable it to deceive the black-box detector.

For performance evaluation, the authors downloaded data from a public website, which contains a repository of malware and benign program instances. A total of 180,000programs were downloaded from the website out of which 30percentwere of malicious category. As the IDS is assumed to be a black-box, therefore several ML algorithms were used to replace the black-box in actual experimentation, the algorithms were LR, random forest (RF), decision tree (DT), support vector machine (SVM) and an ensemble of these algorithms, VOTE. Firstly, the experiments were conducted for the case when the MalGAN and the IDS shared the same data set. As expected, the recall of the black-box is significantly reduced on adversarial examples. Table 6.1 illustrates the performance of black-box IDS on training and test set. In second set of experiment, black-box detector and MalGAN were trained on

**TABLE 6.1**
TPR (%) of the Black-Box IDS on Same Training Dataset with Original and Adversarial Examples

| | Training Set | | Test Set | |
|---|---|---|---|---|
| | **Original** | **Adversarial** | **Original** | **Adversarial** |
| RF | 96.61 | 0.19 | 94.37 | 0.17 |
| LR | 91.19 | 0.00 | 91.35 | 0.00 |
| DT | 97.90 | 0.15 | 93.58 | 0.14 |
| SVM | 92.14 | 0.00 | 92.17 | 0.00 |
| MLP | 94.98 | 0.00 | 93.90 | 0.00 |
| VOTE | 96.45 | 0.00 | 94.78 | 0.00 |

**TABLE 6.2**
TPR (%) of the Black-Box IDS on Different Training Dataset with Original and Adversarial Examples

| | Training Set | | Test Set | |
|---|---|---|---|---|
| | **Original** | **Adversarial** | **Original** | **Adversarial** |
| RF | 94.11 | 0.69 | 93.57 | 0.76 |
| LR | 90.57 | 0.00 | 90.82 | 0.05 |
| DT | 92.17 | 2.15 | 90.27 | 2.19 |
| SVM | 91.35 | 0.00 | 91.76 | 0.00 |
| MLP | 93.56 | 0.00 | 92.78 | 0.00 |
| VOTE | 93.27 | 0.00 | 92.56 | 0.00 |

separate training data set. It is validated by the experiments that the substitute detector faces difficulty to copy the black-box IDS trained on a different training dataset.

Therefore, reduction in true positive rate is comparatively lower than the previous case were both the black-box detector and MalGAN were trained on same training dataset. Table 6.2 displays the result obtained in the second set.

## 6.5 Conclusion

The chapter discussed the application of GANs in the field of designing intrusion detection systems. It can be observed from the research literature that the majority of the work in the area is primarily focused on: (1) use of GANs for addressing data imbalance;(2) application of GANs in classification and (3) using GANs to generate adversarial examples. Section 6.2 discussed various ways in which GANs can be employed to mitigate the data imbalance problem, which can further improve the detection accuracy of malicious traffic by machine learning/deep learning classifier. In Section 6.3 we discussed various research works where the researchers have directly used GAN models for classification task; also the use of GAN is not limited to binary classification and researchers have proposed several ways in which it can even be used for multiclass classification. Section 6.4 discussed some of the research work devoted to the application of GANs for generating adversarial examples for IDS, which can be used to verify robustness of a target intrusion detection system.

## Glossary of Terms, Acronyms and Abbreviations

| | |
|---|---|
| AC-GAN | auxiliary classifier GAN |
| ADASYN | adaptive synthetic |
| CPS | cyber physical systems |
| CSSAE | cost-sensitive stacked autoencoder |
| CW-GAN | conditional Wasserstein GAN |
| DT | decision tree |
| FFNN | feedforward neural network |
| FSC-GAN | few shot classifier GAN |
| GAN | generative adversarial network |
| IDS | intrusion detection system |
| IoT | Internet of Things |
| LR | linear regression |
| MFC-GAN | multiple fake class generative adversarial network |
| MLP | multilayer perceptron |
| NB | naive Bayes |
| RF | random forest |
| ROS | random oversampling |
| SMOTE | synthetic minority oversampling technique |
| SVM | support vector machine |
| TACGAN | tabular auxiliary classifier GAN |

## References

1. Lee, J., and Park, K.. (2021)."GAN-based imbalanced data intrusion detectionsystem."*Personal and Ubiquitous Computing* 25.1: 121–128.
2. Shahriar, M. H., Haque, N. I., Rahman, M. A., andAlonso, M. (2020, July). G-ids: Generative adversarial networks assisted intrusion detection system. In *2020 IEEE 44th Annual Computers, Software, and Applications Conference (COMPSAC)* (pp. 376–385). IEEE.
3. Ali-Gombe, A., and Elyan, E. (2019)."MFC-GAN: class-imbalanced dataset classification using multiple fake class generative adversarial network."*Neurocomputing* 361: 212–221.
4. Zhang, G., et al. (2020)."Network intrusion detection based on conditional Wassersteingenerative adversarial network and cost-sensitive stacked autoencoder."*IEEE Access* 8:190431–190447.
5. Ding, H., et al.(2022)."Imbalanced data classification: A KNN and generative adversarial networks-based hybrid approach for intrusion detection."*Future Generation Computer Systems* 131: 240–254.
6. Li, D., Chen, D., Jin, B., Shi, L., Goh, J., andNg, S. K. (2019, September). MAD-GAN: Multivariate anomaly detection for time series data with generative adversarial networks. In *International Conference on Artificial Neural Networks* (pp. 703–716). Springer.
7. De Araujo-Filho, P. F., et al.(2020)."Intrusion detection for cyber–physical systems usinggenerative adversarial networks in fog environment."*IEEE Internet of Things Journal* 8.8: 6247–6256.
8. Zenati, H., Romain, M., Foo, C.-S., Lecouat, B. and Chandrasekhar, V. (2018). "Adversarially learned anomaly detection," in *Proc. IEEE Int. Conf. Data Min* (pp. 727–736).ICDM.

9. Wang, C., Zhang,Y.-M. and Liu, C.-L.. (2018)."Anomaly detection via minimum likelihood generative adversarial networks." In *2018 24th International Conference on Pattern Recognition (ICPR)* (pp. 1121–1126). IEEE.
10. Yin, C., et al.(2020)."Enhancing network intrusion detection classifiers using supervised adversarial training."*The Journal of Supercomputing* 76.9: 6690–6719.
11. Kanakarajan, N. K., andMuniasamy, K. (2016). Improving the accuracy of intrusion detection using gar-forest with feature selection. In *Proceedings of the 4th International Conference on Frontiers in Intelligent Computing: Theory and Applications (FICTA) 2015* (pp. 539–547). Springer.
12. Ingre, B., andYadav, A. (2015, January). Performance analysis of NSL-KDD dataset using ANN. In *2015 International Conference on Signal Processing and Communication Engineering Systems* (pp. 92–96). IEEE.
13. Wei, Peng, et al.(2019)."An optimization method for intrusion detection classification model based on deep belief network."*IEEE Access* 7: 87593–87605.
14. Potluri, S., Ahmed, S., andDiedrich, C. (2018, December). Convolutional neural networks for multi-class intrusion detection system. In *International Conference on Mining Intelligence and Knowledge Exploration* (pp. 225–238). Springer.
15. Javaid, Ahmad, et al. (2016)."A deep learning approach for network intrusion detection system."*Eai Endorsed Transactions on Security and Safety* 3.9: e2.
16. Lin, Zilong, Shi,Yong, and Xue, Z. (2018)."Generative adversarial networks for attack generation against intrusion detection."*CoRR Abs* 2.1: 12–18.
17. Hu, W., and Tan, Y.(2017)."Generating adversarial malware examples for black-box attacks based on GAN."*arXiv preprint arXiv:1702.05983.*

# 7

# *Textual Description to Facial Image Generation*

**Vatsal Khandor, Naitik Rathod, Yash Goda, Nemil Shah, and Ramchandra Mangrulkar**

**CONTENTS**



## 7.1 Introduction

Deep learning is a sub-branch of machine learning that contains networks skilled in learning unsupervised from unstructured and unlabeled data. Development in deep learning began back in 1943 with the publication of the paper "A Logical Calculus of the Ideas Immanent in Nervous Activity" [1]. This paper contains the mathematical model of biological neurons. Soon after this publication, inspired by this, Frank Rosenblatt develops the Perceptron and also publishes the paper "The Perceptron: A Perceiving and Recognizing Automation" [2]. Deep learning had been primarily used for only trivial tasks of conducting classification and regression. Soon after the formation of perceptron, Marvin Minsky and Seymour Papert published a book *Perceptrons* [3] undermining the works of

DOI: 10.1201/9781003203964-7

Rosenblatt that the perceptron was not capable of solving the complicated XOR function. But as advancements carried out in the domain of deep learning with the introduction of neural networks, the flaws of perceptron got eliminated and a boost has been evidenced in the development in perceptron and the field of deep learning. Deep learning is now capable of performing various multimodal tasks such as text-to-image generation and recognition, and audio-visual speech recognition. Along with these, due to the formation of the ImageNet dataset by Fei-Fei Li et al., 2009 [4] which contains 14 million label images. Also, the evolution of deep learning in the domain of visual content formation boosted due to the introduction of GANs.

Text-to-image synthesis was introduced by Reed in 2016 [5]. This is the fundamental and innovative research field in computer vision. Text-to-image works on similar lines as that of image caption. It helps to search the interrelation between text and image, inspecting the visual semantic process of the human brain. Besides this, it also has huge potential in image searching, computer-aided design, and so on. Text-to-image synthesis uses a trained text encoder to conceal the input descriptions as a semantic vector and then train a conditional GAN as an image decoder to produce realistic images based on the combination of the noise vector and the semantic vector. This synthesis breaks the training process of the image decoder and text encoder. Text-to-face synthesis is the sub-field of the text-to-image synthesis. It seeks the formation of face images based on text descriptions. Text-to-face synthesis objective is similar to text-to-image synthesis of generating high-quality images and generating images accurately according to the input textual descriptions. It handles the evolution of cross-modal generation and multimodal learning and shows great capability in computer-aided design, cross-modal information retrieval, and photo editing.

Humans can draw a mental image immediately as they hear or read a story visualizing the information in their head. This natural ability to visualize and recognize the convoluted relationship between the visual world and the language is hardly acknowledged. Designing a visual mental picture plays a vital role in many cognitive processes such as "hearing with the mind's ear" and "seeing with the mind's eye" [6]. Influenced by how humans envision events, modeling a technique that acknowledges the correlation between vision and language. And on understanding, this correlation, forming an image betraying the meaning of textual descriptions is a significant milestone achieved by Artificial Intelligence. To create the visual contents more user-friendly, the latest developments have been dominating various intermediary modalities, for example, semantic labels [7], textual description as cited in Nam et al. (2018) [8].

In the past few years, there have been immense advancements in the domain of computer vision and image processing techniques. Image synthesis is one of the fields that includes the generation of realistic images using machine learning algorithms as cited in Gopta et al. [9]. Image synthesis is a very crucial task as it contains many practical applications such as image editing, video games, virtual reality, and computer-aided designs. The images produced from these techniques were of low quality. Many problems have occurred for obtaining high-resolution and high-quality images. High-resolution images contain lots of facial minute details like freckles, skin pores, etc. Another problem was a poor generalization. Due to the recent advancements in GANs, a whole different image generation prototype that attains remarkable quality, and realism has been evidence. Generative adversarial network (GAN) is the most eminent model of deep learning for the generation of various images as in Goodfellow et al. [10]. GAN creates new data samples that correspond to the training dataset. For example, GANs can form images that look similar to human facial photographs, but this image doesn't need to belong to a particular person. It is an unsupervised method of learning. GANs learn the patterns and the regularities of the input data to generate new similar data points that draw inference from the training data. They designed the image synthesis task as a two-player game of two engaging artificial neural networks, that is, it consists of two models: generator and discriminator. The generator creates new data points that are similar to the original training data points whereas the discriminator tries to identify whether a given data point is real or fake. Both generator and discriminator try to outdo each other. The generator network's training motive is to raise the error rate of the discriminator's network. At the beginning of training, the generator network produces random images. So, the discriminator can easily identify the fake images. However, as training continues, the images trained start getting more realistic. As the generator begins producing more realistic images, the discriminator finds it difficult to distinguish the fake images from the real images. The training continues until the discriminator fails to distinguish half the images generated by the generator as fake. Applications of GAN have been escalated swiftly. GAN technology has been utilized in the domain of science where it helps in attaining improved and precise astronomical images of the celestial bodies. Also, it has been used for some concerning malicious activities such as deep fakes. Deep fakes referred to the production of fake and incriminating videos and photographs. StyleGAN is one of the highly outstanding GAN frameworks introduced by Nvidia researchers in December 2018 [11]. StyleGAN relies on CUDA software of Nvidia and Tensorflow by Google and GPUs. StyleGAN technology works similar to deep fakes and is used mostly for sinister purposes.

## 7.2 Literature Review

In the past few decades, deep learning has made leaps and bounds of progress. In the past decade, state-of-the-art deep learning architectures have achieved almost human-like accuracy [12]. Apart from classification and regression tasks, deep learning is also used for complex tasks like object detection [13], lip reading [14] as well as sound generation [15]. Active research has been ongoing in the field of multimodal learning too. Various tasks like facial generation from speech [16], pose generation from text [17], etc. Efforts are being made to generate appropriate faces from textual descriptions. For this task, the description needs to be encoded into vectors. Techniques like CharCNNRNN [18], skip-thought networks [19], as well as modified language representation models like SentenceBERT [20].

Since the advent of GANs [21], various GAN architectures have been used for synthesis of images. Researchers have been constantly modifying different architectures of GANs to perform various tasks. Some of the architectures previously used by the researchers include DCGAN [22], conditional GANs [23], progressive GANs as well as StackedGANs [24,25].

DCGANs are a novel architecture that uses deep convolutional networks for unsupervised learning tasks. As convolutional neural networks have been widely successful in performing supervised learning tasks related to computer vision, DCGAN uses these layers to successfully perform image generation tasks too.

Stack GANs consist of various GANs stacked progressively to generate images of superior qualities. These GANs are arranged in a top-down manner. Each GAN tries to generate plausible lower-level representations from the higher-level representations of images. This architecture allows GANs to receive labels and then generate images from them.

Progressively growing GANs, also known as ProGANs [26], is a type of generative adversarial network that makes use of a training approach that grows progressively. Both the generator and discriminator models are trained progressively: training is first done at lower resolutions first, while new layers are added successively to ensure that the model learns finer details with progress in training.

## 7.3 Dataset Description

For performing the determined task, the Text2FaceGAN dataset is used. This dataset is a work of students and collaborators at MIDAS Research as cited in Nasir et al. [27].

This dataset is based on the CelebA dataset. It is a dataset of natural language descriptions of human faces. The dataset includes crowdsourced annotated descriptions of faces of various celebrities, which have been verified by the dataset creators to expunge objectionable as well as irrelevant descriptions. Although the dataset was primarily created for the generation of accurate descriptions from facial images, the availability of acute descriptions as well as eclectic faces makes it a perfect fit for text-to-image generation tasks too.

Overall, the CelebA dataset [28] contains

- 202,599 facial images of various celebrities
- 40 binary attribute annotations per image for example black hair, brown hair, wears earrings, etc.

The Text2FaceGAN dataset has added captions to each of these images from CelebA dataset using GAN. Multiple captions are provided under each of the images so that using all the captions under one image; accurate images can be obtained using the descriptions for the Text2Face tasks. The size of the images in the datasets is 178×218 pixels.

Dataset preprocessing: The procured dataset was resized into 64 * 64 pixels image as this is the size of the images required for DCGAN. Along with this, the CenterCrop transform is applied to the image. The CenterCrop transform crops the image from the center. As most of the important features of the dataset are located at the center of the image, this ensures that the most important features of the dataset are preserved while resizing. DCGAN also recommends normalization of images before training. Then the dataset is normalized as [0.5,0.5,0.5] is used as the mean whereas [0.5, 0.5, 0.5] is the standard deviation of the dataset. This preprocessing ensures smooth training of the GAN.

## 7.4 Proposed Methodology

In the past few decades, various advancements have been made in the field of deep learning. Deep learning techniques are now widely used for multimodal learning. Multimodal learning involves more than one channel of data like speech, text, and images. One of the major multimodal tasks is the visual representation of text. Facial image generation from text is one of the major Text2Image (T2I) research problems. The novel architecture of GANs makes them an appropriate fit for generative tasks. So, GANs have been proposed for the task of generating realistic images of faces from the text. The architecture of GANs has been proposed below:

Generative adversarial networks (GANs) were first proposed in [10]. GANs comprise two main components: (i) generator and (ii) discriminator. As suggested by the name, the main aim of the generator is to generate new data points similar to the training inputs whereas the discriminator tries to determine if the image is real or created by the generator. The aim of the discriminator is to correctly predict whether the image was synthesized by the generator whereas the generator aims to fool the discriminator to incorrectly classify its images. Thus, both the discriminator and generator are involved in a minmax game which is represented by

$$\min_{G} \max_{D} V(D,G) = \mathbb{E}_{x \sim P_{\text{data}}(x)}\left[\log D(x)\right] + \mathbb{E}_{z \sim P_z(z)}\left[\log\left(1 - D(G(z))\right)\right] \tag{7.1}$$

In this equation,

G represents generator,
D represents discriminator.

Equation (7.1) represents the interaction between the generator and the discriminator of the GAN.

This innovative approach of making two neural networks compete against each other instead of stacking them together in an ensemble makes GAN truly unique. For generating images of faces, conditional DCGAN is used, which is explained in [24]. In the conditional GAN, the labels of the images are also encoded and provided as inputs along with the images. This helps the GAN to produce desired images. DCGAN [29] is one of the widely used GANs for generating highly realistic images. They use only convolutional layers in the generator and the discriminator, which make them computationally efficient. The various constraints for training DCGANs make the training of GANs quite stable. Various researchers have tweaked the architecture of DCGANs to make computationally expensive GANs, which produce high-resolution images. It consists of deconvolutional as well as convolutional layers and can generate images of size 64 * 64 pixels. The architecture of both the generator and discriminator has been explained further.

### 7.4.1 Generator

The main aim of the generator is to produce realistic images based on the input text. For this process, the textual embeddings are first converted to a 512-dimensional vector. For this process, universal sentence encoder (USE) is used. Figure 7.1 displays the intermediate steps carried out in the generator.

The USE is one of the state-of-the-art document embedding techniques. It is created by Google researchers and available publicly in TensorFlow-hub. Document embedding involves representation of a sequence of words in an $n$-dimensional vector. Document embedding can be considered as extension of word embeddings. USE can be used as a pre-trained model to encode text into 512 dimensional vectors for conversion of sentences to embeddings.

Embedding techniques like FastText and Word2Vec are only useful for word-level implementation. They are not useful when it comes to applying embeddings on sentences.

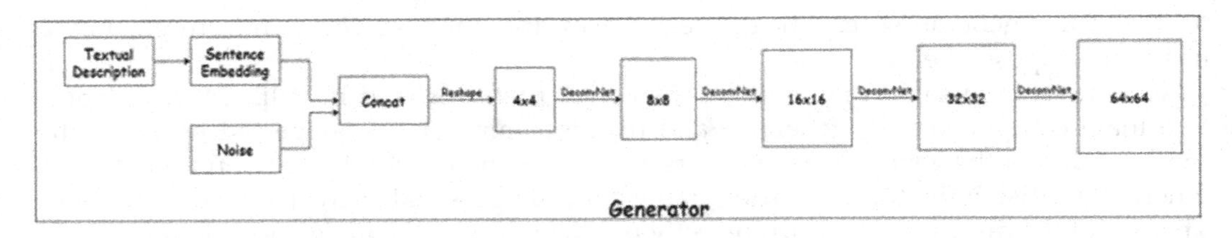


**FIGURE 7.1**
Generator architecture.

**Source: [22]**

USE has two different architectures, which produce efficient accurate embeddings from the user provided input. Although both the architectures produce accurate embeddings, each architecture has some peculiarities. The DAN (deep averaging architecture) network prioritizes efficiency over accuracy and is widely used for producing computationally efficient embeddings. The transformer architecture is complex and computationally expensive. However, it produces very accurate embeddings, which can be used in tasks, which require very accurate descriptions. These inputs can include a phrase, single sentence, verse, an essay or even an entire book. These embeddings are later used in a variety of tasks ranging from sentence classification, clustering to text similarity [30]. These embeddings can also be provided as an input to other models for a variety of other natural language modelling (NLM) tasks.

The two variations for the Universal Sentence Encoder are shown in Figure 7.2:

Figure 7.2 shows both, the USE transformer architecture, as well as the DAN architecture.

#### *7.4.1.1 DAN (Deep Averaging Network)*

As the name suggests, the input embeddings of the input words and bi-grams will be averaged and sent through a feedforward neural network to produce sentence embeddings. The encoder here takes a tokenized string as an input and the output is a sentence embedding containing 512 dimensions as explained in [30]. At the final layer, a softmax activation function is applied. A softmax activation function is used to calculate the probability distribution from a vector of real numbers.

$$\sigma(\vec{z})_i = \frac{e^{z_i}}{\sum_{j=1}^{K} e^{z_j}} \tag{7.2}$$

Equation (7.2) represents the softmax activation function.

The compositionality of the inputs is learned by the neural network and it is observed that it makes errors similar to syntactically aware models. Even though the model is simplistic, it outperforms many complex models, which are designed explicitly to learn the compositionality of the text input.

There are two types of composition functions:

- Syntactic (considers order of words)
- Unordered (a simple bag of words)

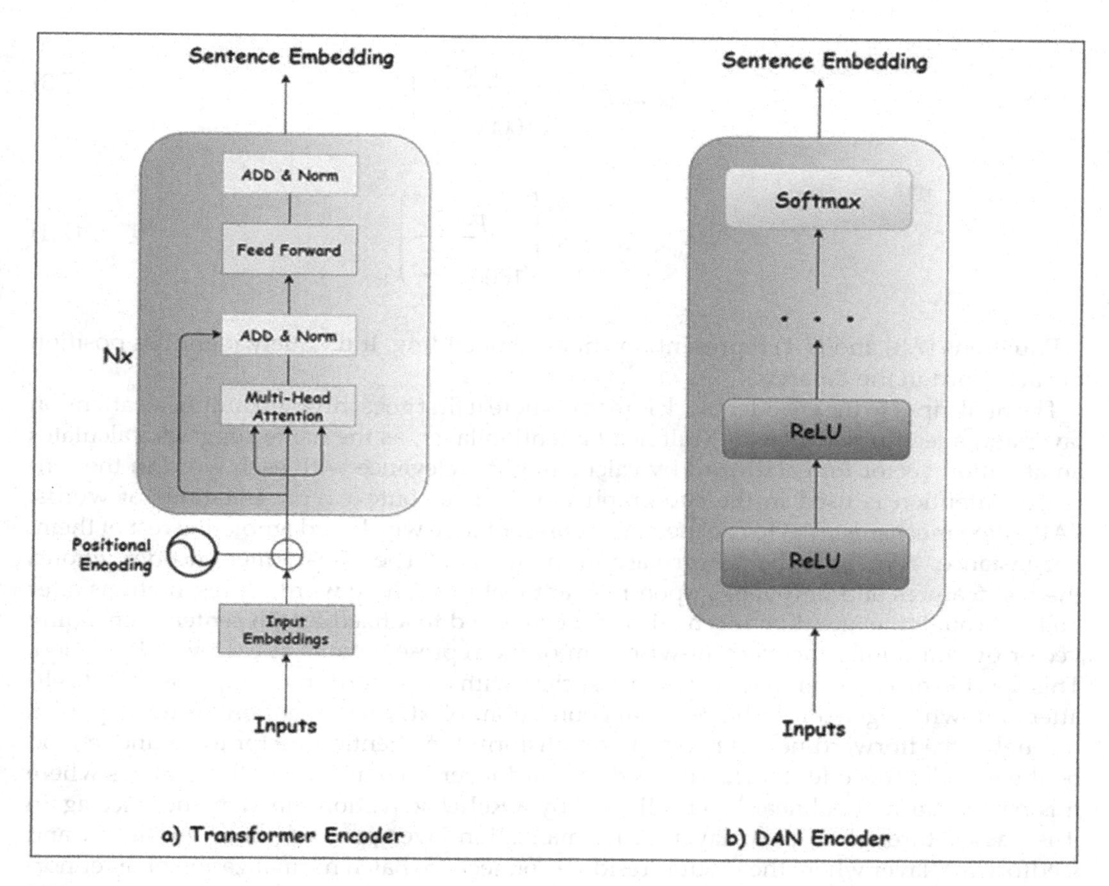


**FIGURE 7.2**
Universal sentence encoder architecture.

**Source:** [30]

This DAN approach [31] takes comparatively lesser time and is computationally less expensive and gives a little less accuracy than the other variation, that is, the transformer encoder.

#### *7.4.1.2 Transformer Encoder*

Transformer employs the encoder–decoder architecture much like the recurrent neural networks (RNN). Here, the input sequence can be passed in parallel. Unlike RNN, in which the word embeddings are produced one after the other, here the word embeddings for the entire sentence are calculated simultaneously.

It uses the sub-graph of a transformer architecture [32], that is, the encoder part. The architecture is composed of six stacked transformer layers. All the layers in the sub-graph have a self-attention module.

It utilizes the embedding space where words having similar meanings are grouped and assigned a particular value. After receiving this encoding, positional embedding is added so the new vector obtained is based on the position of the word in the sentence since same word might have different contexts in different sentences. This step is crucial because the vector generated is now also based on context.

$$PE_{(pos,2i)} = \sin\left(\frac{pos}{10000^{\frac{2i}{d_{model}}}}\right) \tag{7.3}$$

$$PE_{(pos,2i+1)} = \cos\left(\frac{pos}{10000^{\frac{2i}{d_{model}}}}\right) \tag{7.4}$$

Equations (7.3) and (7.4) represent positional embedding. It uniquely identifies position of each word in the dataset.

The final input to the encoder block is ready where it first goes through multihead attention layer and a feedforward layer. Multihead attention layer, as the name suggests, calculates an attention vector for each word by calculating its relevance with each word in the sentence. Attention is used in the sub-graph to calculate context representations of words. "Attention mechanism" is to use the main representative words and ignore the rest of them. For instance, "Where is the flower vase in the image?" The transformer encoder ignores the rest features and only looks upon relevant features. These word representations after going through the attention mechanism are converted to a fixed length sentence encoding vector by computing the element-wise sum of the representations at each word position. This specific mechanism that allows to associate with each word in the input is called self-attention, which generates the residual connection. Next, each attention vector is passed through a feedforward neural network to transform the attention vector to be understood be the encoder/decoder block. The feed forward layer is composed of three stages where it is passed through a linear layer followed by a ReLU activation function and once again it is passed through a linear layer. A normalization layer proceeds both multihead and feedforward layer where the input is residual connection. Batch normalization or layer normalization may be utilized to make the training easier and stable. This encoder block can be stacked multiple times to encode more and thus each layer can learn the different attention representations which aids the predictive power of the transformer network.

So, it can be concluded that the transformer encoder takes a tokenized string as an input and the output is a sentence embedding containing a 512-dimensional vector similar to the deep averaging network (DAN).

This has higher accuracy than DAN, but at the same time, it is more computationally expensive. This variation of USE can handle a single word, sequence of words, or even a document as input.

Though computationally expensive, the focus in text to face is maximizing the accuracy and thus a transformer encoder is preferred over the deep averaging network.

The 512-dimensional sentence embedding is compressed to a 128-dimensional vector to improve computational speed. This is done by passing the embedding through a fully connected neural network and then using the ReLU activation function. In a fully connected neural network, all the neurons in a single layer are connected to each neuron of the next layer. ReLU (Rectified Linear Unit) is an activation function that returns the input if the input is positive, else it returns zero.

This 128-dimensional embedding is then concatenated with a 100-dimensional noise vector. The noise vector has a mean of 0 and 0.2 standard deviation. This concatenated vector is then reshaped into a size of 4*4. These vectors are then passed through a series of block, which consists of 2D fractionally strided convolutional layer [33] (also known as deconvolutional layer or 2D convolutional transpose layer), along with 2D BatchNorm [34] and Leaky ReLU [35]. At each step, the number of filters is reduced by half whereas the new size of the image is twice the original size. The output of these blocks is then passed through a tanh activation to produce the final output. The final image has a size of 64* 64 pixels and three channels where each channel represents 1 of the RGB colors. All these layers are explained below:

*2D convolutional transpose layer*: This layer is used to increase the size of input, that is, upsampling the images. For DCGANs, a kernel size of four is used along with a stride of two to produce images of double the size.

*2D Batch Norm*: Batch normalization is one of the most effective regularization techniques. In batch normalization, each input is first subtracted by the mean and divided by the standard deviation of the batch. Then it is offset by a learned scaled mean and standard deviation for better convergence.

*Leaky ReLU*: Leaky ReLU is one of the most popular activation functions. It solves the vanishing gradients problems caused by ReLU activation as ReLU outputs 0 if the input is negative. In LeakyReLU, the output is the same as ReLU when the input is positive; however, it outputs a small gradient when the inputs are non-zero. This makes it differentiable even for negative inputs.

*Tanh*: The hyperbolic tan function is applied before the final output. It is used as it can project the inputs in the range (–1,1).

### 7.4.2 Discriminator

The aim of the discriminator is to predict whether the images are ground truth or were created by the generator. However, apart from this it also needs to predict whether the image and the text are mismatched or not. Such a discriminator is also known as matching aware discriminator. This is essential as although the discriminator can easily distinguish between the images at the start of the training process; once the generator starts producing realistic images, the discriminator might need to check if the generated image aligns with the textual description to produce better images. For this purpose, three types of inputs are provided to the discriminator. The discriminator is provided with the real image along with the correct text pair as well as the generator output along with its correct textual description. However, it is also provided with mismatched pairs, that is, real images along with incorrect textual descriptions. This allows the discriminator to correctly identify if the images actually represent their descriptions correctly and thereby produce even better images. The components of the discriminator are quite similar to those of the generator. A 64 * 64 image is provided as the input along with the description as displayed in Figure 7.3. Then the image is passed through a series of blocks containing a convolutional layer, batch normalization layer as well as LeakyReLU activation function. Convolutional layers drastically reduce the computational time and resources used for image segmentation tasks. A convolutional layer is mainly used for the task of downsampling the images. A convolutional layer performs the convolution operation, that is, conversion of all the pixels in the receptive field to a single output. The convolutional layer consists of a set of learnable filters where each filter outputs a set of parameters that adequately represent the receptive field. A 2D convolutional layer with a kernel of size 4 and a stride of 2 is used in the discriminator. The series of blocks convert the 64 * 64 image to a 4*4 block. This block is then flattened and concatenated with the 128-dimensional embedding. Thais is then passed through the sigmoid function which predicts whether the image is created by the generator. The sigmoid activation function is a nonlinear activation function which has a S-shaped curve. The sigmoid function has a range of (0, 1). It is widely used for binary

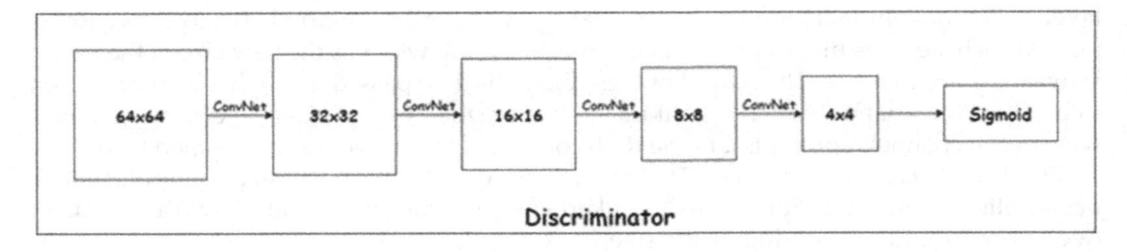


**FIGURE 7.3**
Discriminator architecture.

**Source:** [22]

classification. If the output of the sigmoid is close to 0 it is fake, that is, image is generated by the generator, or the text doesn't match the image. If the output is 1, the image is real and matches the textual description.

### 7.4.3 Training of GAN

The generator and discriminator of a GAN are trained simultaneously. First, the discriminator is trained on the random untrained images of the generator. Then the training of the discriminator is stopped, and the generator is trained on the learned parameters of the discriminator. Gradually, the generator improves the quality of images and successfully fools the discriminator. However, the images of the generator are still not realistic. So, the discriminator is again trained on the newer images created by the generator. This process continues until the GAN produces realistic images. The training of GAN is quite tricky, and several methods have been proposed for the smooth training of GAN. These factors are described as follows:

#### *7.4.3.1 Loss Function*

For the loss function to train the GAN, the BCELogits Loss [36] should be used. It combines both the sigmoid layer and the binary CrossEntropy loss in one single class.

$$l(x,y) = \{l_1,\ldots,l_N\}^T, l_n = -w_n\left[y_n \cdot \log\sigma(x_n) + (1-y_n)\cdot\log(1-\sigma(x_n))\right] \quad (7.5)$$

Equation (7.5) explains the loss function of the GAN training.

The BCELogits Loss provides more stability than using the sigmoid followed by the BCELoss. By combining both the operations in a single layer, the log-sum-exp trick can be used which results in the numerical stability.

#### *7.4.3.2 Optimizer*

Adam as cited in Kingma et al. [37], Adam optimizer is one of the most effective and robust optimization techniques used in deep learning. The use of Adam optimizer has been found to produce state-of-the-art results while training GANs. Adam uses mini batch gradient descent with momentum, that is, aggregation of past gradients to accelerate convergence along with aggregation of the squares of past gradients like RMSProp. Adam makes use of exponential moving averages to roughly estimate both the momentum as well as the second moment of the gradient. Adam effectively combines all the properties of other optimization algorithms.

#### *7.4.3.3 Discriminative Learning Rates*

Both the discriminator and the generator must use different learning rates [38]. Usually, the learning rate of the discriminator is higher than that of the generator. As the generator needs to learn the latent representations of the dataset to produce similar images, the lower learning rate in the generator leads to greater stability. However, the discriminator can easily learn the differences and a higher learning rate would save a lot of computational time and resources.

#### *7.4.3.4 Dropout*

As proposed in Isola et al. [39], use of 50 percent dropout in various layers of the generator during both the training as well as testing phase improves the stability and results of GAN. Dropout [40] is the technique of purposefully injecting noise into the internal layers of the network. As this technique literally drops out some neurons, it has been named as dropout. The use of dropout has improved accuracy of various state-of-the-art models. In dropout, each neuron has a probability $p$ of being dropped out of the process. This probability factor is known as dropout rate. As neurons trained on dropout need to pay attention to inputs of all the neurons as any neuron can be dropped out of the network, dropout results in better stability.

## 7.5 Limitations

The amount of accuracy attained of the generated image is not as expected due to the availability of a finite amount of data pertaining to facial images. The description of the facial image should contain a generation to create similarity between the facial image generated and the textual description related to it. The suggested approach produces facial images of size 64×64. It fails to generate images of size 128×128 and 256×256 due to availability of limited computational power.

This chapter's current technique is limited only for generating facial images. This concise the development related to this topic. Also this technique doesn't ensure complete and robust security to the user information. Along with these, this approach requires lengthy textual description for the generation of the facial images instead of some particular keywords, which can help in generating the similar facial images as obtained from the extensive textual description.

## 7.6 Future Scope

Currently a very limited amount of data is available for the facial images, so the datasets can be improvised further. Increased datasets can play a vital role in achieving greater accuracies in the deployed models. The descriptions of the facial images can be subjective from person to person, so a framework for facial description can be generalized which would help to increase the similarity between the textual description and the facial image generated. Generalized framework could be fruitful in the employment of text2face models in the public safety domain.

Existing systems can only produce images that are 64×64 in size while better output images can be achieved using higher computational power. The task of super-resolution can also be implemented in the image generated. The current models are restricted to facial images only, so there is huge scope of improvement here in the future. A system can be developed which considers the entire images of people that is, finding images of the body that match the descriptions provided. This can be very beneficial in public safety. Future work can be done in the field relating to whole body image generation using the textual descriptions.

In future, even systems can be designed in such a way that specific keywords could be inputted to the GUI and the system will construct an image that will match the key terms. There will be no need to write big descriptions for fetching the images and only key words will do the job.

## 7.7 Conclusion

Deep learning was initially used for the image classification, language translation, automation tasks, etc. With the introduction of neural networks, the field of deep learning is now capable of performing various multimodal tasks such as text-to-image generation and audio-visual speech recognition. GAN is the most used in the deep learning domain when tasks involving images as GAN creates new images based on the present dataset which helps in better classification of the text. The authors here proposed an approach using GAN to generate facial images using the textual description provided. Although computationally expensive, the transformer encoder is used over the deep averaging network in the sentence encoding to achieve greater accuracy. The sentence embedding is compressed into a 128-dimensional vector using a fully connected neural network and then using ReLU. The output of this block is then fed into tanh function for the final sentence embeddings. Matching aware discriminator checks if the image is generated by the generator or is it a true image. It also checks if the image and the text matches or not. These two are the main tasks of the discriminator. The training of the generator and the discriminator is done simultaneously. Multiple factors have to be considered while training GAN like loss functions, optimization using Adam, discriminative learning rates of the generator and the discriminator and the use of dropouts for increasing stability and results of GAN.

The text-to-face generation has huge potential in domains like public safety. It is hoped the proposed methodology in this chapter could be beneficial for this task.

## 7.8 Applications

This technique can be implemented for criminal identification and security purposes. It helps in the automatic face generation of criminals using the machine learning approach

(GAN). With the textual information provided by the eyewitness, it can draw out a facial sketch. This would in turn help the authorities in identifying and arresting the criminal and providing justice to the victim. Furthermore, this technique would help in text-to-image translation. Besides this, it also aids in generating sketches of original faces from the provided textual description. Also, this helps in translating visual concepts from characters to pixels using the GAN architecture.

It would also help in visualization of faces as determined by the plastic surgery patients as it would provide them with the virtual display of their faces based on the textual description presented by them so that they can acquire the desired facial alterations they wish to model their face. There are sometimes shortages of facial datasets like there may be a lack of data or there may be an issue regarding the privacy of the user. So in this scenario, this technique will help in creating datasets containing facial data and it will also ensure that the privacy of the users is preserved and also form a good quality of dataset.

## References


1. McCulloch, Warren S., and Walter Pitts. "A logical calculus of the ideas immanent in nervous activity." *The Bulletin of Mathematical Biophysics* 5, no. 4 (1943): 115–133. https://doi.org/10.2307/2268029
2. Rosenblatt, Frank. "The perceptron: a probabilistic model for information storage and organization in the brain." *Psychological Review* 65.6 (1958): 386. https://doi.org/10.1037/h0042519
3. Minsky, Marvin, and Seymour A. Papert. *Perceptrons: An Introduction to Computational Geometry*. MIT Press, 2017. https://doi.org/10.7551/mitpress/11301.001.0001
4. Fei-Fei, Li, Jia Deng, and Kai Li. "ImageNet: Constructing a large-scale image database." *Journal of Vision* 9, no. 8 (2009): 1037–1037. https://doi.org/10.1167/9.8.1037
5. Reed, Scott, Zeynep Akata, Xinchen Yan, Lajanugen Logeswaran, Bernt Schiele, and Honglak Lee. "Generative adversarial text to image synthesis." In *International Conference on Machine Learning*, pp. 1060–1069. PMLR, 2016. http://proceedings.mlr.press/v48/reed16.html
6. Kosslyn, Stephen M., Giorgio Ganis, and William L. Thompson. "Neural foundations of imagery." *Nature Reviews Neuroscience* 2, no. 9 (2001): 635–642. www.nature.com/articles/35090055
7. Isola, Phillip, Jun-Yan Zhu, Tinghui Zhou, and Alexei A. Efros. "Image-to-image translation with conditional adversarial networks." In *Proceedings of the IEEE Conference on Computer Vision and Pattern Recognition*, pp. 1125–1134. 2017. https://doi.org/10.1109/CVPR.2017.632
8. Nam, Seonghyeon, Yunji Kim, and Seon Joo Kim. "Text-adaptive generative adversarial networks: manipulatqing images with natural language." *arXiv preprint arXiv:1810.11919* (2018). https://arxiv.org/pdf/1810.11919.pdf
9. Gupta, Param, and Shipra Shukla. "Image synthesis using machine learning techniques." In *International Conference on Intelligent Data Communication Technologies and Internet of Things*, pp. 311–318. Springer, Cham, 2019. https://doi.org/10.1007/978-3-030-34080-3_35
10. Goodfellow, Ian J., Jean Pouget-Abadie, Mehdi Mirza, Bing Xu, David Warde-Farley, Sherjil Ozair, Aaron Courville, and Yoshua Bengio. "Generative adversarial networks." *arXiv preprint arXiv:1406.2661* (2014). https://arxiv.org/abs/1406.2661
11. Abdal, Rameen, Yipeng Qin, and Peter Wonka. "Image2stylegan: How to embed images into the stylegan latent space?." In Proceedings of the IEEE/CVF International Conference on Computer Vision, pp. 4432–4441. 2019. https://arxiv.org/abs/1904.03189

12. Karpathy, Andrej. "What I learned from competing against a ConvNet on ImageNet." Andrej Karpathy Blog 5 (2014): 1–15. http://karpathy.github.io/2014/09/02/what-i-learned-from-competing-against-a-convnet-on-imagenet/
13. Redmon, Joseph, and Ali Farhadi. "Yolov3: An incremental improvement." *arXiv preprint arXiv:1804.02767* (2018). https://arxiv.org/abs/1804.02767
14. Assael, Yannis M., Brendan Shillingford, Shimon Whiteson, and Nando De Freitas. "Lipnet: End-to-end sentence-level lipreading." *arXiv preprint arXiv:1611.01599* (2016). https://arxiv.org/abs/1611.01599
15. Oord, Aaron van den, Sander Dieleman, Heiga Zen, Karen Simonyan, Oriol Vinyals, Alex Graves, Nal Kalchbrenner, Andrew Senior, and Koray Kavukcuoglu. "Wavenet: A generative model for raw audio." *arXiv preprint arXiv:1609.03499* (2016). https://arxiv.org/abs/1609.03499
16. Oh, Tae-Hyun, Tali Dekel, Changil Kim, Inbar Mosseri, William T. Freeman, Michael Rubinstein, and Wojciech Matusik. "Speech2face: Learning the face behind a voice." In Proceedings of the IEEE/CVF Conference on Computer Vision and Pattern Recognition, pp. 7539–7548. 2019. https://doi.org/10.1109/CVPR.2019.00772
17. Ahuja, Chaitanya, and Louis-Philippe Morency. "Language2pose: Natural language grounded pose forecasting." In 2019 International Conference on 3D Vision (3DV), pp. 719–728. IEEE, 2019. https://doi.org/10.1109/3DV.2019.00084
18. Zhang, Xiang, Junbo Zhao, and Yann LeCun. "Character-level convolutional networks for text classification." *arXiv preprint arXiv:1509.01626* (2015). https://arxiv.org/abs/1509.01626
19. Kiros, Ryan, Yukun Zhu, Ruslan Salakhutdinov, Richard S. Zemel, Antonio Torralba, Raquel Urtasun, and Sanja Fidler. "Skip-thought vectors." *arXiv preprint arXiv:1506.06726* (2015). https://arxiv.org/abs/1506.06726
20. Reimers, Nils, and Iryna Gurevych. "Sentence-bert: Sentence embeddings using Siamese bert-networks." *arXiv preprint arXiv:1908.10084* (2019). https://arxiv.org/abs/1908.10084
21. Creswell, Antonia, Tom White, Vincent Dumoulin, Kai Arulkumaran, Biswa Sengupta, and Anil A. Bharath. "Generative adversarial networks: An overview." *IEEE Signal Processing Magazine* 35, no. 1 (2018): 53–65. https://doi.org/10.1109/MSP.2017.2765202
22. Fang, Wei, Feihong Zhang, Victor S. Sheng, and Yewen Ding. "A method for improving CNN-based image recognition using DCGAN." *Computers, Materials and Continua* 57, no. 1 (2018): 167–178. https://doi.org/10.32604/cmc.2018.02356
23. Dai, Bo, Sanja Fidler, Raquel Urtasun, and Dahua Lin. "Towards diverse and natural image descriptions via a conditional gan." In Proceedings of the IEEE International Conference on Computer Vision, pp. 2970–2979. 2017. https://arxiv.org/abs/1703.06029
24. Zhang, Han, Tao Xu, Hongsheng Li, Shaoting Zhang, Xiaogang Wang, Xiaolei Huang, and Dimitris N. Metaxas. "Stackgan: Text to photo-realistic image synthesis with stacked generative adversarial networks." In Proceedings of the IEEE International Conference on Computer Vision, pp. 5907–5915. 2017. https://arxiv.org/abs/1612.03242
25. Huang, Xun, Yixuan Li, Omid Poursaeed, John Hopcroft, and Serge Belongie. "Stacked generative adversarial networks." In Proceedings of the IEEE Conference on Computer Vision and Pattern Recognition, pp. 5077–5086. 2017. https://arxiv.org/abs/1612.04357
26. Gao, Hongchang, Jian Pei, and Heng Huang. "Progan: Network embedding via proximity generative adversarial network." In Proceedings of the 25th ACM SIGKDD International Conference on Knowledge Discovery & Data Mining, pp. 1308–1316. 2019. https://doi.org/10.1145/3292500.3330866
27. Nasir, Osaid Rehman, Shailesh Kumar Jha, Manraj Singh Grover, Yi Yu, Ajit Kumar, and Rajiv Ratn Shah. "Text2FaceGAN: Face Generation from Fine Grained Textual Descriptions." In 2019 IEEE Fifth International Conference on Multimedia Big Data (BigMM), pp. 58–67. IEEE, 2019. https://doi.org/10.1109/BigMM.2019.00-42
28. Liu, Ziwei, Ping Luo, Xiaogang Wang, and Xiaoou Tang. "Deep learning face attributes in the wild." In Proceedings of the IEEE International Conference on Computer Vision, pp. 3730–3738. 2015. https://doi.org/10.1109/ICCV.2015.425

29. Radford, Alec, Luke Metz, and Soumith Chintala. "Unsupervised representation learning with deep convolutional generative adversarial networks." *arXiv preprint arXiv:1511.06434* (2015). https://arxiv.org/pdf/1511.06434.pdf
30. Cera, Daniel, Yinfei Yanga, Sheng-yi Konga, Nan Huaa, Nicole Limtiacob, Rhomni St Johna, Noah Constanta et al. "Universal sentence encoder for English." *EMNLP 2018* (2018): 169. www.aclweb.org/anthology/D18-2029.pdf
31. Minaee, Shervin, Nal Kalchbrenner, Erik Cambria, Narjes Nikzad, Meysam Chenaghlu, and Jianfeng Gao. "Deep learning-based text classification: A comprehensive review." *ACM Computing Surveys (CSUR)* 54, no. 3 (2021): 1–40. https://doi.org/10.1145/3439726
32. Vaswani, Ashish, Noam Shazeer, Niki Parmar, Jakob Uszkoreit, Llion Jones, Aidan N. Gomez, Lukasz Kaiser, and Illia Polosukhin. "Attention is all you need." *arXiv preprint arXiv:1706.03762* (2017). https://arxiv.org/abs/1706.03762
33. Dumoulin, Vincent, and Francesco Visin. "A guide to convolution arithmetic for deep learning." *arXiv preprint arXiv:1603.07285* (2016). https://arxiv.org/abs/1603.07285
34. Ioffe, Sergey, and Christian Szegedy. "Batch normalization: Accelerating deep network training by reducing internal covariate shift." In *International Conference on Machine Learning*, pp. 448–456. PMLR, 2015. https://proceedings.mlr.press/v37/ioffe15.html
35. Xu, Bing, Naiyan Wang, Tianqi Chen, and Mu Li. "Empirical evaluation of rectified activations in convolutional network." *arXiv preprint arXiv:1505.00853* (2015). https://arxiv.org/abs/1505.00853
36. BCEWithLogitsLoss—PyTorch Master Documentation'. Accessed May 15, 2021. https://pytorch.org/docs/master/generated/torch.nn.BCEWithLogitsLoss.html
37. Kingma, Diederik P., and Jimmy Ba. "Adam: A method for stochastic optimization." *arXiv preprint arXiv:1412.6980* (2014). https://arxiv.org/abs/1412.6980
38. Howard, Jeremy, and Sebastian Ruder. "Universal language model fine-tuning for text classification." *arXiv preprint arXiv:1801.06146* (2018). https://arxiv.org/abs/1801.06146
39. Isola, Phillip, Jun-Yan Zhu, Tinghui Zhou, and Alexei A. Efros. "Image-to-image translation with conditional adversarial networks." In Proceedings of the IEEE Conference on Computer Vision and Pattern Recognition, pp. 1125–1134. 2017. https://doi.org/10.1109/CVPR.2017.632
40. Srivastava, Nitish, Geoffrey Hinton, Alex Krizhevsky, Ilya Sutskever, and Ruslan Salakhutdinov. "Dropout: a simple way to prevent neural networks from overfitting." *The Journal of Machine Learning Research* 15, no. 1 (2014): 1929–1958. https://jmlr.org/papers/v15/srivastava14a.html

# 8

# *An Application of Generative Adversarial Network in Natural Language Generation*

**Pradnya Borkar, Reena Thakur, and Parul Bhanarkar**

**CONTENTS**



## 8.1 Introduction

Generative adversarial networks (GANs) are a type of generative modelling that employs deep learning techniques such as convolutional neural networks. Generic modelling is an unsupervised machine learning problem that entails automatically detecting and learning regularities or patterns in incoming data so that the model may be used to create or output new instances that could have been chosen from the original dataset. GANs are a rapidly evolving field that fulfils the promise of generative models by generating realistic examples across a wide range of problem domains, most notably in image-to-image translation tasks such as converting summer to winter or day to night photos, as well as generating photo-realistic photos of objects, scenes, and people that even humans can't tell are fake.

GANs are a clever way to train a generative model by framing it as a supervised learning problem with two sub-models: the generator model, which we train to generate new

DOI: 10.1201/9781003203964-8

examples, and the discriminator model, which tries to classify examples as real (from within the domain) or fake (from outside the domain) (generated). The two models are trained in an adversarial zero-sum game until the discriminator model is tricked around half of the time, indicating that the generator model is creating convincing examples.

## 8.2 Generative Adversarial Network Model

The task is framed as supervised learning with two sub-models: the generator model, which we train to generate new instances, and the discriminator model, which tries to categorise cases as true (from your dataset) or false (from your dataset) (generated) (see Figure 8.1).

- *Generator*: The model is used to produce new credible instances from the area of the problem.
- *Discriminator*: The model is used for determining if instances are genuine (from the domain) or not (generated).

The two models are trained in an adversarial zero-sum game until the discriminator model is tricked around half of the time, indicating that the generator model is producing convincing examples.

### 8.2.1 Working of Generative Adversarial Network

The discriminator evaluates whether each instance of data it analyses corresponds to the actual training dataset or not, while the generator generates new data instances. Let's pretend we're attempting something other than a Mona Lisa impersonation. We'll create handwritten numerals based on real-world data, comparable to those seen in the MNIST collection. The discriminator's goal when given an example from the true MNIST collection is to recognise which ones are genuine.

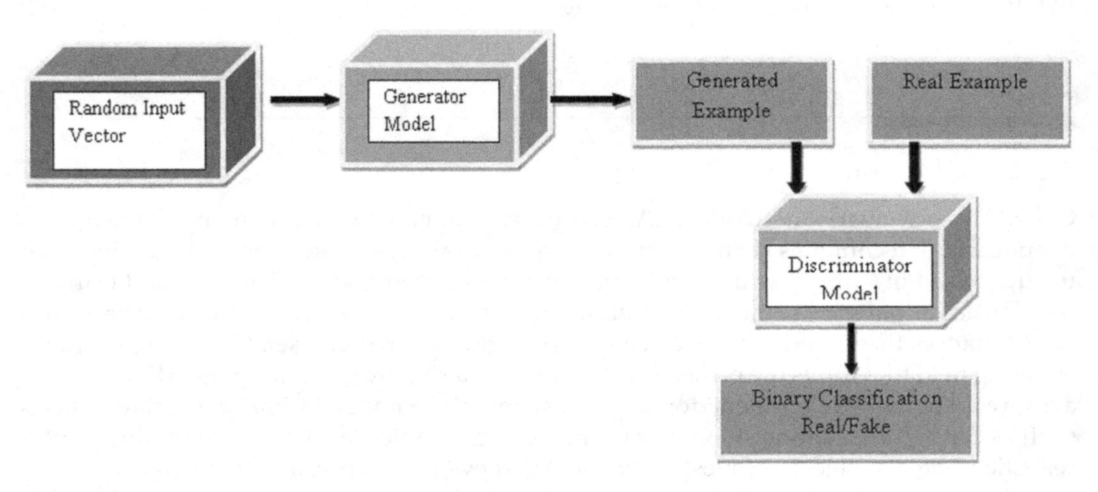


**FIGURE 8.1**
Basic block diagram of GAN model.

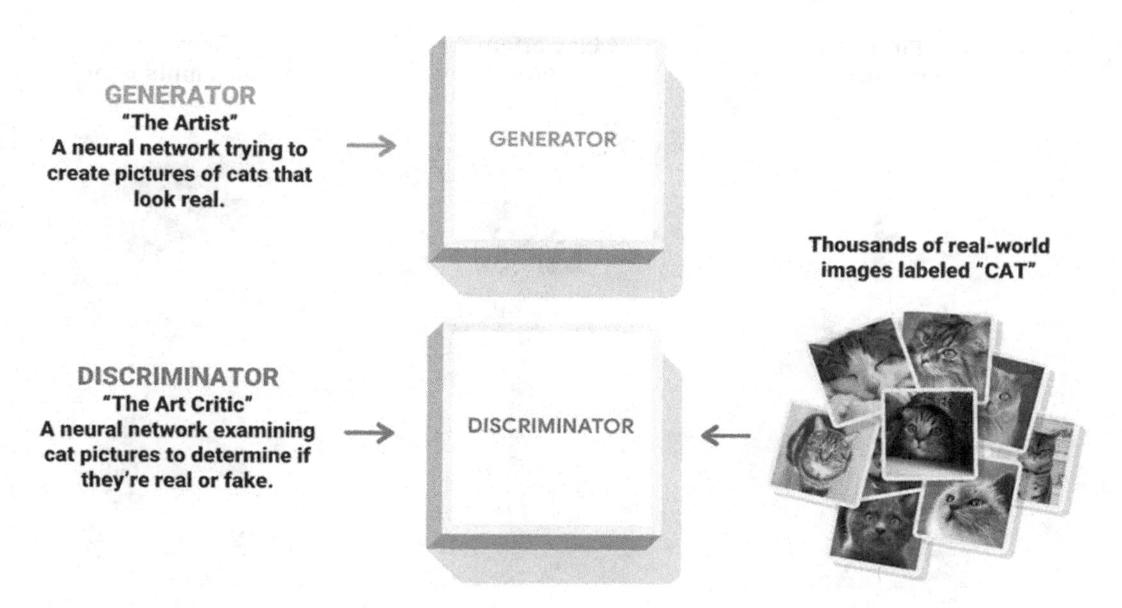


**FIGURE 8.2**
Example of generator and discriminator.

**Source:** [1]

Meanwhile, the discriminator is processing new synthetic images generated by the generator. It does so in the hopes that, despite the fact that they are not, they will be recognised as genuine. The goal of the generator is to generate presentable handwritten digits that allow the user to cheat without being detected. The discriminator's job is to identify bogus images created by the generator.

Some steps of GAN are explained as follows:

- The generator returns an image after taking random numbers.
- The discriminator is fed a stream of photos taken from the actual, ground-truth dataset alongside a stream of images generated by the generator.
- The discriminator returns probabilities by accepting both actual and false photos and, which range from 0 to 1, with 1 indicating authenticity and 0 indicating fake.

An adversarial approach is used to train the two models simultaneously, as shown in Figure 8.2. A generator ("the artist") always learns to create realistic images, but a discriminator ("the art critic") learns to distinguish between actual and fraudulent images.

As shown in Figure 8.3, it shows that while training, the generator becomes better progressively for creating images that look real, whereas the discriminator becomes better at telling them apart. At the certain level when the process reaches one of the equipoise positions, at that time the discriminator cannot differentiate real images from fake images.

### 8.2.2 Natural Language Generation

The process of creating sentences, dialogue, or books without the use of humans is called natural language generation (NLG). It includes both natural language processing (NLP)

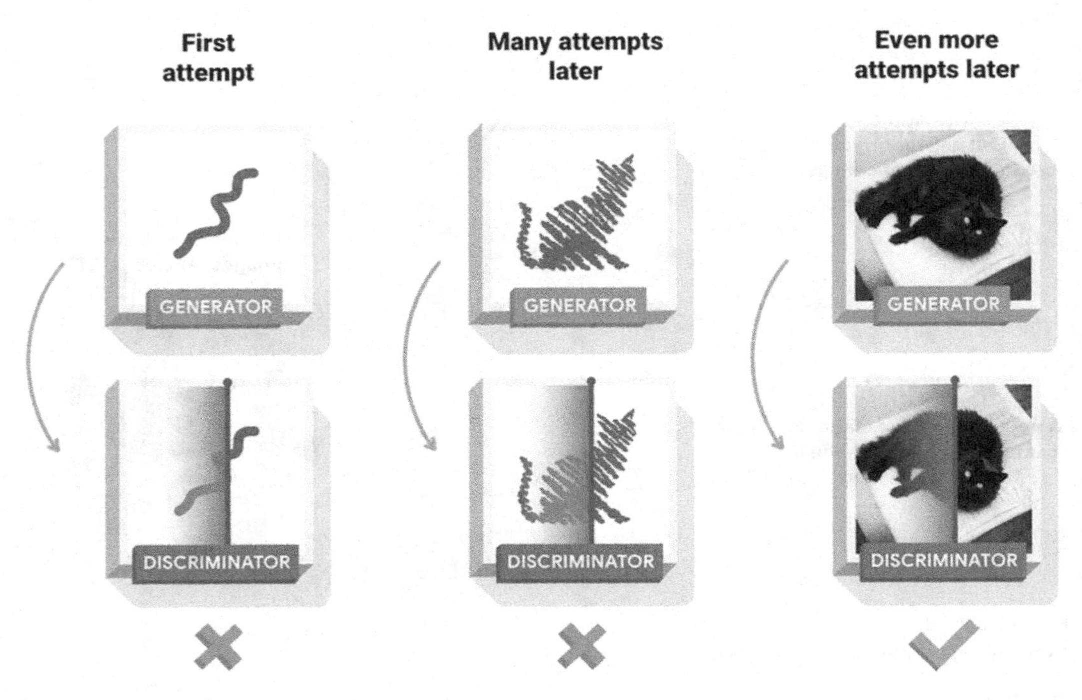


**FIGURE 8.3**
Example (training) of generator and discriminator.

**Source:** [1]

and natural language understanding (NLU), but it is a separate area with its own set of uses and capabilities. NLG has a wide range of applications, from the mundane to the bizarre. The most common application of NLG is in content generation. NLG has already been used to create music lyrics, video game dialogue, and even stories. The results are promising, even if they aren't flawless. They could be enhanced in the future to generate game worlds, create music albums, and even write complete novels.

This type of data synthesis can be applied to non-creative industries as well. For example, creating synthetic medical data for study without jeopardising patients' identities. Another important application is the automated creation of reports. Weather reports, for example, can be generated automatically from the underlying data without the need for human intervention. Inventory and medical reports can also be generated in the same way. One last use case worth mentioning is that of expanding accessibility. It can be utilised by personal assistants like Siri, Cortana, or Google Voice to describe facts and graphs to persons with disabilities. Natural language generation is a relatively recent field, with the majority of breakthroughs occurring in the last two years. This is owing to the recent surge in machine learning, particularly deep learning.

Because of the current surge in deep learning, this is the case. While recurrent neural networks (RNNs) have been employed successfully in the past, they suffer from a number of disadvantages. One of the major disadvantages is the difficulty in evaluating the created language. The primary purpose of neural networks is to complete a single task. The model doesn't know how to evaluate language because the aim is to generate it. Traditional evaluation metrics are ineffective since this is a generational issue. There is no such thing as "accuracy" here. As a result, people must manually analyse the model.

This is a time-consuming and unrealistic task. The authors have been employed a generative adversarial network to solve these disadvantages (GAN). Two networks, a "discriminator" and a "generator", are used in a GAN. The generator works in the same way that traditional RNNs do for language generation. By taking over the process of manual evaluation, the discriminator network solves the problem. It's a network that's been taught to "discriminate" or classify the language generated by the generator network. The main characteristic of GAN is that the two networks involved are not mutually exclusive. They are playing a non-cooperative zero-sum game. This implies that they are in competition with one another. Consider how they work after a few revisions. The generator provides an imitation of natural language with each iteration. After that, the discriminator is shown both the generated text and a real-life example of natural language. However, there is no indication of which is which. It must make a decision. When a forecast is made, it is compared to the actual outcome. If the discriminator made a mistake, it examines why it chose that particular option and learns from it. As a result, it improves its ability to recognise natural language.

The generator, on the other hand, failed to do its job if the discriminator chose the correct option. As a result, the generator learns why the discriminator discovered it and improves. The following iteration begins after this process is completed. As a result, the generator begins a protracted game of attempting to deceive the discriminator. With respect to responsibilities, both the generator and the discriminator improve with time. The ideal situation would be for the generator to produce text that is indistinguishable from natural language. As a result, the discriminator would have a 50/50 chance of selecting either option. However, in practice, this would take far too much time and effort [2].

## 8.3 Background and Motivation

Supervised learning-based models of machine learning are the ones that gives best outcome when trained properly with the labelled inputs. Majority of the machine learning community use supervised learning methods, which help generate more reliable results. A mapping function is needed to map the input variable to the output variable. The problems related to classification and regression falls into the supervised learning problems. The basic classification model considers categories or classes for performing the classification task. A simple classification task can be one where the output variable classifies based on the classes like man or animal over the given input variable. The regression problem is based on the real values as it is a statistical model. There are various types of regression models available which include linear regression, polynomial regression, decision tree regression, random tree regression and many more. The unsupervised models on the other hand does not involve any training over the input data. Also the data available for the processing is unlabelled. These models include clustering, neural networks, etc. The recent development in the field is the design of the generative model, which are based on the deep learning methods. The challenge in developing such a model arises from the idea for a better future by generating results that closely resembles real world. In general, terms generative adversarial networks are the innovation of machine learning statistical models which concentrate on the generations of sample instances that typically resemble the samples from the training data set. The GANs nowadays are forming the active and beneficial statistical based machine learning method, which has the capability to deal with most complex datasets and bring out surprising results.

Over the past decades, the unsupervised pre-training concept and the emergence of deep learning methodologies have unfolded many aspects of new research. This started way back in 1943–2006, where the idea was to train a two-layer unsupervised deep learning model and continue adding, training the layers to improve learning capability. The neural networks then became the deep learning networks, which has now applications in almost every field. They are now being trained to do almost everything that human can do. Also they helped explore the fields that are untouchable for humans. The very crucial component of these networks are the graphics processing units (GPUs), which could be of hundreds to thousands of cores, having the fast processing capabilities and could help reduce the error rate in the models designed. The error issue called as dropout was handled with the help of rectified linear activation unit. ReLU. The work on the deep learning networks improved with the massive high-quality labelled data sets, massive parallel computing, back-propagation activation functions, robust optimisers. GANs are the recent models based on deep learning. The deep learning statistical models can also be generative and discriminative. The generative describes statistical models that actually contrast with the representations of the discriminative models. The generative models are capable of generating new instances from the training data whereas the discriminative models are capable of making discriminations between the data instances generated. For example, the generative models can create new images of birds that look realistic. The discriminative model will be able to discriminate between the type of birds. The generative model work on joint probability while the discriminative models work on the conditional probability. There are certain issues and challenges in the implementation of the GANs, which include initialisation of the model, collapsing of the modes, counting problem etc. The work discusses the various challenges during the implementation of GANs for various applications.

## 8.4 Related Work

In [3], for neural machine translation, the author employed a conditional sequence generative adversarial network (CSGAN-NMT). A generator and a discriminator are the two sub-models in the proposed model. Based on a source language, the generator generates text. And the discriminator assesses this translation by estimating the likelihood that it is the accurate translation. To achieve Nash equilibrium, a gamified mini–max procedure is used between the two sub-models to arrive at a win–win situation. The generator was made out of a gated recurrent units (GRUs) encoder–decoder system with 512 hidden units.

These criteria were chosen to prevent the manually constructed loss function from leading to the development of inferior translations. They used their network on the NIST Chinese–English dataset, and results were also presented on an English–German translation assignment to further assess the approach's efficiency. For the production of translations, a beam search was used. The beam width, for example, was set to ten, and the log-likelihood values were not standardised by sentence length. The models were built in TensorFlow and trained synchronously on up to four K80 GPUs in a multiGPU configuration on a single system. Experiment results using the proposed CSGANNMT model showed significant outperformance as compared to previous efforts. They also explored a variant known as the multi-CSGAN-NMT, which is a scenario in which several generators and discriminators are used to achieve spectacular outcomes, with each generator acting as an agent, interacting with other generator agents, and even sending messages. The generators were able to learn significantly better thanks to the usage of two separate discriminators. Various testings have revealed that the alternate option outperforms the original proposed model even more. They also discovered that discriminators with an accuracy that was either too high or too low fared poorly in their testing.

In [4], the authors have suggested a gated attention neural network model (GANN) with two major components. First, there's a comment generator, which is based on an encoder–decoder framework, with the encoder component converting all words in the title into one-hot vectors and embedding representations can be obtained by multiplying the embedding matrix. The last hidden vector of the title triggers the initialisation of the generator's decoder component. The model, like the encoder, coverts the sequence of comment words into one-hot vectors and uses the shared embedding matrix to obtain their low-dimensional representations. Different weights can be assigned to distinct elements of related information; modules such as the gated attention mechanism and a relevance control module are introduced to ensure relation between comments and news, which has been demonstrated to increase performance. The second element is a comment discriminator, which is used to improve the accuracy of comment generation. This is a concept inspired by GANs. By performing various tests on huge dataset, GAANs show the effectiveness as compared to other generators. The generated news comments were found to be close to human comments. Computational improvements in medical research have been fuelled by the widespread adoption of electronic health record records by healthcare organisations, as well as an increase in the quality and amount of data. However, there are a number of privacy considerations that bounds access to and shared use of this data.

In [5], Choi et al. built the medical generative adversarial network (medGAN) to produce realistic synthetic patient records by providing genuine patient records as input. By combining autoencoder with GAN into design, binary and count variables can be handled. Count variables represent the number of times a specific diagnosis was made, whereas binary variables represent the greatest likelihood estimate of the diagnostic's Bernoulli success probability.

Using a technique known as tiny batch averaging, this method avoids mode collapse and overfitting the model to a smaller number of training samples. TensorFlow was used to create MedGAN with a learning rate of 0.001 and a mini-batch of 1000 patients, as well as Adam optimiser. The Sutter Palo Alto Medical Foundation dataset, the MIMIC-III dataset, and a heart failure study dataset from Sutter have all been used to test MedGAN. Furthermore, by including the reporting of distribution statistics, classification performance, and medical expert assessment, medGAN gives close-to-real data performance. medGAN does not simply recall and recreate the training samples, according to privacy experiments. Rather, medGAN creates a variety of synthetic samples that tell very little about the intended patient.

In [6], the authors have suggested an adversarial training method for generating realistic text. It uses the traditional GAN architecture of a generator and a discriminator, with a long short-term memory network (LSTM) as the generator and a convolutional network as the discriminator. This approach proposes a matching between high-dimensional latent feature distributions real and synthetic phrases rather than the traditional GAN, which is accomplished using a kernelised discrepancy measure. The suggested framework components solve the mode-collapsing problem, making adversarial training easier.

When compared to other relevant approaches, this particular model outperforms them. It not only generates believable words, but it also allows the learnt latent representation space to smoothly encode them. The methods were quantitatively tested using baseline models and existing methods as benchmarks, and the results show that the above-proposed methods outperform them.

The technique of creating distractors for a set of fill-in-the-blank questions and keys to the question is known as distractor generation. It is widely used to discriminate between test takers who are knowledgeable and those who are not. In [7], the author proposes a conditional GAN model for distractor generation that relies solely on stem information and current key-based approaches. Distractors are generated in this model to better fit the query context.

This machine learning-based method differs from earlier efforts in that it is the first to use a GAN for distractor generation. The extra context learnt from the question set and key is used to condition both the discriminator and the generator. The generator takes a joint input of an input noise vector and the context, while the discriminator assesses if the generated sample comes from the training data.

A two-layer perceptron with a hidden size of 350 is used to create the GAN model. Leaky ReLU is the activation function employed in this model. Each LSTM has a hidden size of 150 and uses the Adam algorithm, which has a learning rate of 0.001 during training. The training data consists of a lexicon of biology-related themes with a total of 1.62 million questions and answers. The Wiki-FITB dataset and the Course-FITB dataset were used to evaluate the model. On the same data, the GAN model was compared to a similarity-based Word2Vec model, as well as the evaluation process. The Word2Vec model was found to be limiting because it could only generate the same distractors regardless of question stems, whereas the GAN model could generate a variety of distractors. In order to obtain better performance, the GAN and Word2Vec models were coupled to minimise the percentages of problematic distractors while increasing the diversity of the distractors. In the future, a unified GAN model will be included, which may be used for better context modelling and model generalisation.

## 8.5 Issues and Challenges

To train a GAN, two networks, the generator and the discriminator, must compete against one other to find an optimal solution, more specifically a Nash equilibrium. The definition of Nash equilibrium is a stable condition of a system containing numerous persons' interactions in which no single member may unilaterally change their strategy if the strategies of the others remain unchanged.

1. *Failure and faulty initialisation setup*: Both the generator and the discriminator reach a point where they can no longer improve while the other remains constant. The goal of gradient descent is now to reduce the loss measure defined on the problem—but in GAN, which has a non-convex objective with continuous high-dimensional parameters, the goal is to reduce the loss measure defined on the problem. We are not requiring the networks to reach Nash equilibrium. The networks attempt to minimise a non-convex objective in a series of stages, but instead of diminishing the underlying actual objective, they end up in an oscillating process. You can usually

tell immediately away if something is wrong with your model when your discriminator reaches a loss extremely close to zero. Finding out what's wrong is the most challenging aspect. Another effective GAN training strategy is to cause one of the networks to stall or learn at a slower rate so that the other network can catch up. Because the generator is usually the one that is late, we usually let the discriminator wait. This may be acceptable to some extent, but keep in mind that for the generator to improve, a good discriminator is essential, and vice versa. In a perfect scenario, the system would like both networks to learn at a rate that permits them to keep up with each other.

2. *Modes collapse*: Another useful technique used during GAN training is to deliberately cause to allow the other network to catch up, One of the networks must stall or slow down its learning. We normally let the discriminator wait because the generator is usually the one who is late. This may be acceptable to some extent, but keep in mind that a good discriminator is required for the generator to develop, and vice versa. Both networks should learn at a rate that allows them to improve over time in an ideal environment, according to the system. The discriminator's optimal minimal loss is close to 0.5, meaning that from the discriminator's perspective, the generated images are indistinguishable from the originals.
3. *Problem with counting*: GANs can be farsighted at times, failing to distinguish the quantity of certain objects that should occur at a given place. As can be seen, it results in a greater number of eyes in the head than were initially present. When it comes to training GAN models for execution, it is critical, and there are certain frequent issues that may arise. The failure of the setup provides a serious difficulty, and mode collapse, also known as the helvetica scenario, is a regular occurrence when training GAN models. It clarifies a number of topics, including counting, viewpoint, and global organisation.
4. *Instability*: The instability is the problem related to the unstable training issues in the GAN. This generally relates with the synchronisation of the generator and the discriminator involved. If the performance balance for both is not maintained, the model may become unstable. The model generated may sometime seem to be stable, but the developed model may not be able to generate the stable output. The variations of the GAN-like KL-GAN could be used to improve over the said issue. The deep convolutional GAN can be used to reduce the unstable training issues.
5. *Non-convergence*: The GAN model in many cases fails to converge during the training of the model. In this issue, the model parameters destabilise and are not able to meet the expected outcomes. This can also cause due to the imbalance that can be seen between the generator and the discriminator. One of the solutions could be avoiding the overfitting during their training. The cost functions can be improved to avoid the non-convergence.
6. *Training time of the model*: The detailed exploration of the generation of the input space makes the model slower to respond. For example, the generator when needs to generate the natural sentences, it needs to explore various combinations of words to reach at the expected outcome. The drop out ratio and the batch size affects the learning rate due to which it is difficult to find the better model.
7. *Diminished gradients*: The problem of vanishing gradients occurs when the training is given to the model with the help of gradient based methods. In this case, the gradient selected may be sometimes very small and vanishing. An optimal discriminator

should be able to provide information to make better progress of the model. To address the problem, a modified minimax loss function can be developed. The vanishing gradients can also be managed with the help of Wassertein loss.

## 8.6 Case Studies: Application of Generative Adversarial Network

GANs have made a drastic change in the computer vision tasks. A variety of fields could be benefitted with the advantages provided by the GAN networks. Out of the many applications of the GAN natural language processing has gained greater attention [9]. The applications in trend include data augmentation, translations over images like cityscapes to the night time, painting to photograph, sketch to photograph, etc. Text-to-image conversion, face view generation and many more.

### 8.6.1 Creating Machines to Paint, Write, Compose, and Play

- *Write*: Generating text, tokenisation, and how RNN extensions build on top of generative networks. LSTM layer.
- *Compose*: Using generative deep learning to create music, recurrent neural networks and the Music GAN generator.
- *Play*: The use of generative deep learning in reinforcement learning for videogames. How the reinforcement learning systems that implement these algos are trained and used. Variational Auto Encoders.
- *The future of generative modelling*: creation of images, music, and text using generative deep learning for art. Transformers with BERT and GPT-2 being covered.

### 8.6.2 Use of GAN in Text Generation

RNNs are used in deep learning networks to generate text in traditional methods. RNN variants such as LSTMs and GRUs fall within this category. A GAN tuned to text data is presented in this section. This GAN uses two neural networks, a generator and a discriminator, to implement the aforementioned RNN versions. In order to learn from each other, the two networks engage in a zero-sum non-cooperative game. For the creation of the model described here, a systematic and iterative approach was adopted. While the basic system architecture remains constant, each network's core specification differs based on tests. Although the model can be trained on any text, it is preferable to use shorter texts because larger texts take longer to train. It should be easy to train it on any big text corpus given adequate processing power and time.

*Dataset*: When it comes to machine learning activities, the data must generally be in a specific format and separated into a "training" and "testing" set. This is due to the fact that most of these jobs entail predicting a variable. This is not true, however, when it comes to the creation of language. There isn't a single variable that can be predicted. Language, on the other hand, is simply generated. Any corpus of text can be utilised as an input dataset with this in mind. In order to create our model, we limited ourselves to a few well-known datasets. It should be noted, however, that the model can be applied to any corpus of text.

*System architecture*: It consists of just three modules as shown in Figure 8.4. The preprocessing module handles cleaning of the input data and converts it into a machine-friendly representation. The generator network is responsible for attempting to generate text while the discriminator network judges the text.

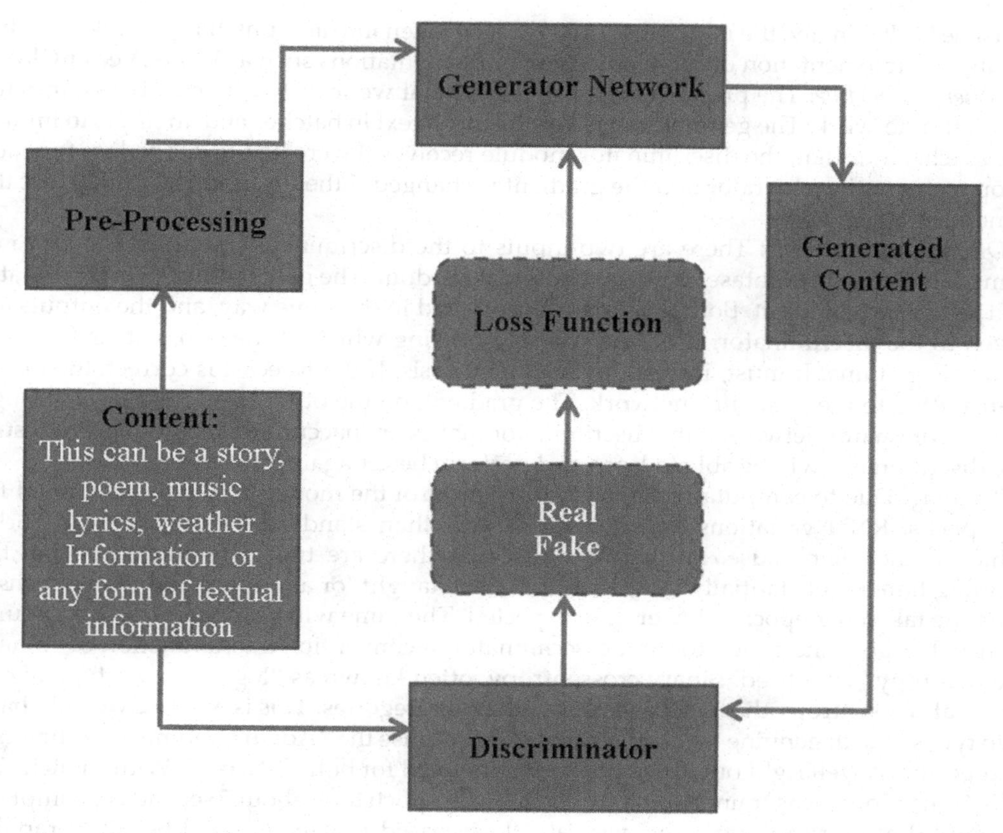


**FIGURE 8.4**
System architecture.

The loss function will propagate and update either the discriminator network or the generator network based on the network's output. Each network will learn more and more over time, resulting in even greater results. As a result, the model will be able to succeed and possibly even deceive humans.

*Preprocessing module*: Rather than including larger words, it is easier for a model to identify tiny numerical patterns in order to generate text. That example, a model can quickly comprehend a series of numbers. It is incapable of comprehending arbitrary text strings. Because computers work in binary rather than text, all text must be represented as numbers. However, there are some obstacles to overcome. We won't be able to just turn the text into numbers. The first thing to think about is the case. Is "Word" synonymous with "word"? It is, although a naive approach would classify them as two distinct terms. As a result, all words must be lowercased. The text's phrases are the next thing to consider. First and foremost, sentences must be detected. There isn't always a period at the end of a sentence. Some have a question mark at the end, while others have an exclamation mark.

Finally, not every sentence is created equal. Some will be brief, while others will be lengthy. However, a model cannot accept such unbalanced data. As a result, we must determine a good standard sentence length that is neither too long, requiring most sentences to be padded, nor too short, requiring the majority of sentences to be trimmed. Regardless of the length of the sentence, both will be required. As a result, functions to perform the same must be built. Finally, the real embedding must be taken into account. It is possible to utilise a simple implementation or more advanced implementations such as Word2Vec or GloVe.

*Generator module*: The preprocessed text corpus that we wish to replicate is sent into the generator network. The generator receives the input text in batches and attempts to imitate the batch. After that, the discriminator module receives the created text. The loss function propagates to the generator and the gradient is changed if the discriminator finds that the generated text is fake.

*Discriminator module*: There are two inputs to the discriminator module. The first is a sample text from the dataset that was chosen at random. The generator network generates text as the second input. Both texts are preprocessed in the same way, and the outputs are given to the discriminator. It has no way of knowing which of the two texts is fake and which is genuine. It must, instead, make a prognosis. If the forecast is correct, the loss is transmitted to the generator network. The gradient, on the other hand, will pass through the discriminator network if the discriminator makes an inaccurate prediction. As a result, the discriminator will be able to learn and perform better against future samples.

*Training*: Due to computational constraints, most of the models will only be trained for 10 epochs. RNN variations are typically slower than standard feedforward networks, which is the main cause. Furthermore, because there are two networks in action, the training time is substantially increased. It may be taught for a longer number of epochs if the time taken per epoch is lower than expected. The same will be noted in such a circumstance. For both the generator and discriminator, a conventional loss function of "binary crossentropy" is utilised. Binary crossentropy, often known as "log loss", is a type of categorical crossentropy that only operates with two categories. This is because we only have two types of text: genuine text and created text. Because the "Adam" optimiser is universal and performs well right out of the box, it is employed for both networks. With a batch size of 512, the model was trained for 100 epochs. Each epoch took about 5 seconds to complete. Around the 64-epoch mark, the generator loss ceased to decrease and began to rapidly increase. The discriminator loss, on the other hand, stayed in the same range of 0.54. The discriminator accuracy, on the other hand, continuously improved until it reached 90% [2].

### 8.6.3 Indian Sign Language Generation Using Sentence Processing and Generative Adversarial Networks

In this case study, GANs are applied for the generation of the Indian sign language. Sentence preprocessing is done over the input sentences, converting them into notations of glosses type and then getting the video frames for the gloss. These glosses are represented using a skeletal representation. This representation is then used as input to the GAN model. The GAN generates video frames and then a video from the same is produced. The technique applied here is the computer vision based model that generated sign video for the input word sequence. This model is useful for the differently abled person to interpret the given information in a helpful way. Quality sign video can be generated for an unknown base given image and some given words. The future techniques can be applied for deploying it into mobile settings and other platforms irrespective of the locations. Figure 8.5 depicts the general GAN structure model for above application [8].

### 8.6.4 Applications of GAN in Natural Language Processing

GANs have made a drastic change in the computer vision tasks. A variety of fields could be benefitted with the advantages provided by the GAN networks. Out of the many

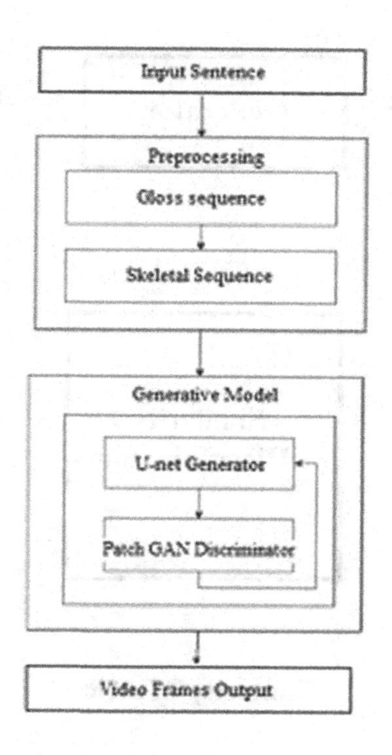


**FIGURE 8.5**
General structure model depicts the general GAN structure model for Indian sign language generation using sentence processing application.

applications of the GAN natural language processing has gained greater attention. The applications in trend include data augmentation, translations over images like cityscapes to the night time, painting to photograph, sketch to photograph, etc. Text-to-image conversion, face view generation and many more.

1. Creating Machines to Paint, Write, Compose, and Play
   - *Write*: generating text, tokenisation, and how RNN extensions build on top of generative networks. LSTM layer
   - *Compose*: using generative deep learning to create music, Recurrent Neural Networks and the MusicGAN generator.
   - *Play*: The use of generative deep learning in reinforcement learning for videogames. How the reinforcement learning systems that implement these algos are trained and used. Variational Auto Encoders.
   - *The future of generative modelling*: creation of images, music, and text using generative deep learning for art. Transformers with BERT and GPT-2 being covered.

GANs find applications in the field of image super resolution and processing, semantic segmentation including face ageing, natural language processing, time series data generation, to name a few. In this part of the work, applications of GANs in the natural language processing is discussed. A few sub-fields of natural language processing include text synthesis, texture synthesis, speech synthesis, lyrics synthesis, and sign language synthesis.

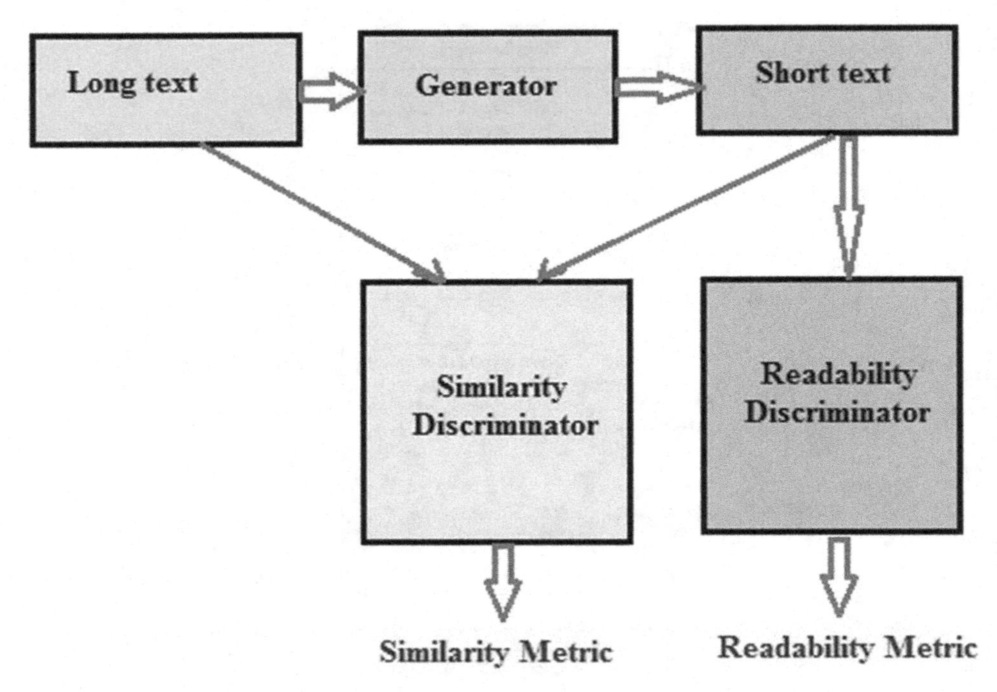


**FIGURE 8.6**
Abstractive text summarisation model.

These represent the extra ordinary ability of the GANs, which are also capable of handling detailed level of metrics of the data.

- Abstractive Text Summarisation [10]:

The abstractive text summarisation concentrates to train the GAN to generate the text that is human readable over the input text source. The summary that is generated is called the abstractive text summarisation. This text summary generated should be semantically equal. Figure 8.6 shows the idea.

The main components are the generator unit that interprets the long text given in the input and converts into short words, the discriminator unit takes responsibility of creating the abstract summary that closely relates to the original text. The policy gradient algorithm can be used to optimise the model here.

- Generating Synthetic Electronic Medical Record Text [11]:

The advancements made in the field of natural language processing and the application of machine learning in the medical domain has led to find new alternatives for the data synthesis. The huge amount of the free-text electronic medical records (EMR) is in research nowadays due to lack of data analysis methods and lack of the availability. The generation of the synthetic EMR text is a challenge that can be addressed with the help of GANs along with reinforcement learning algorithm. The issues that can be handled also includes the generation of better and variety of EMR texts and more realistic ERM text outputs in order to help understand the diseases in study.

- Stimulus Speech Decoding [12]

Decoding of the auditory stimulus from the neural activity could be used to provide direct communication with the brain. This type of speech decoding can be done using the deep learning models. The GANs can be utilised to develop deep learning models in order to decode the intelligible audio stimuli from the input ECoG recordings obtained for the cortical regions of the brain. The only issue that is to be dealt is the limited training samples. The representation space for the training can be extended with the help of the standard Gaussian distribution. The developed model is called as the transfer GAN.

- Text to Image Synthesis from Text Semantics [13]

The text-to-image synthesis has a variety of applications in the area of natural language processing and the image processing. In the text-to-image synthesis, the process includes analysing the text and trying to create a realistic image utilising the text description. Computer-aided design is the most practical application of the same where it can be used to better the design. GANs can work better in utilising the complete semantic information from the given text and try generating the images that are more robust in nature. While generating the semantics, the algorithm learns text features from the description that are more important visual details. When this is done, these features are then used to synthesise the image. The pattern of attention GAN could be used for generation of coarse and fine-grained images.

- Multidimensional Text Style Transfer [14]

The style of text transfer in the field of natural language processing has been in research in the recent times. Text style transfer is the branch of style transfer methods where the objective is to transform the original input text into different desired style. The application of such methods can be in the area of sentiment transformation, formality modification. For implementing text style transfer, the unified generative adversarial networks (UGANs) can be utilised. The UGAN works as an efficient and flexible unified structure for implementing style of text transfer as the model allows training of multi-attribute data using a single generator and single discriminator. The multidirectional text style transfer performed in this manner gives a good performance of the model.

- Text Summary Generation [15]

The GAN is used along with the bidirectional encoder representation from transformers can be used for the automatic text summary generation. The model can be used in combination with the leak GAN to output the generated abstracts. The leak GAN model as shown in Figure 8.7 applies hierarchical reinforcement learning to generate a better and accurate summary. The figure shows the process of leak GAN.

The leak GAN has the capability to solve the generative automatic summarisation problems. The discriminator adds attention mechanism for concentrating on the useful information relating to the input. The convergence speed of the model is good for text applications.

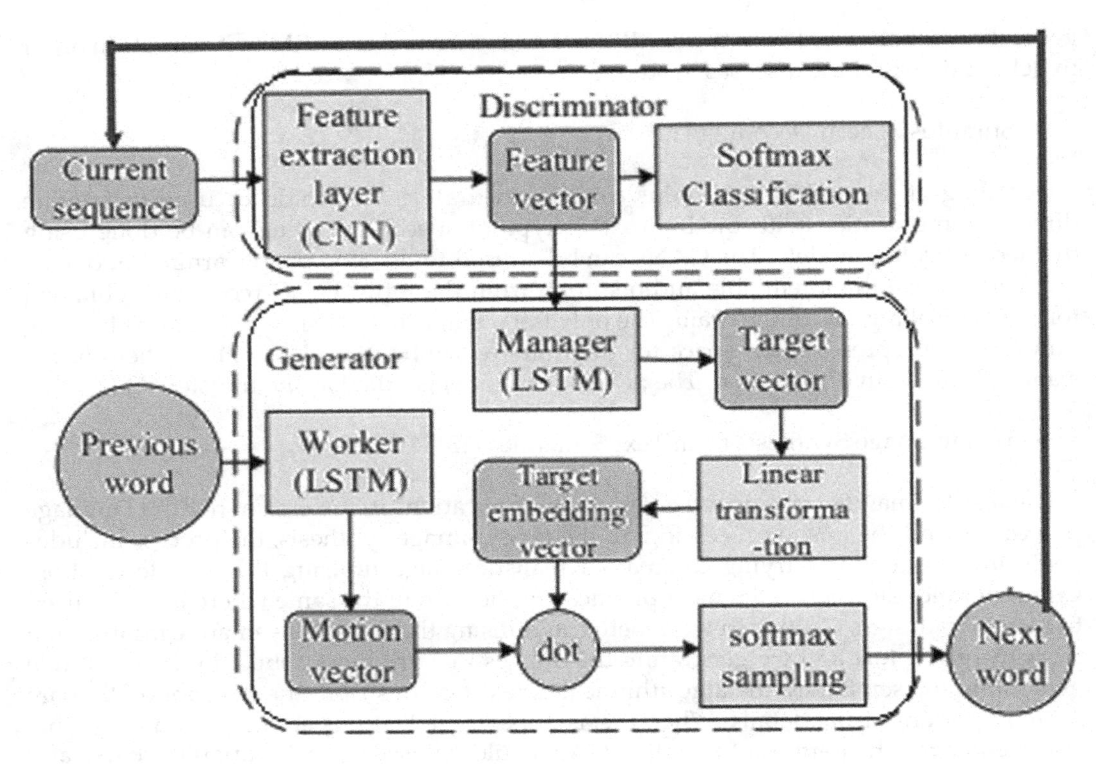


**FIGURE 8.7**
Leak GAN architecture.

## 8.7 Conclusions

GANs are a sort of generative modelling that uses convolutional neural networks and other deep learning techniques. This includes automatic finding and learning of particular kind of patterns in the input set that is used to produce or give new examples as output. GAN can be used in the natural language generation in the natural language generation and text summarisation. The various challenges, methodologies, and applications that includes the use of GAN (generative adversarial network) were discussed.

## References

1. www.tensorflow.org/tutorials/generative/dcgan
2. Saravanan, Akash & Perichetla, Gurudutt & Menon, Akhil & S, Sanoj. (2019). Natural Language Generation using Generative Adversarial Networks. 10.13140/RG.2.2.35169.45922.
3. Yang, Z., Chen, W., Wang, F., and Xu, B. (2018). "Generative Adversarial Training for Neural Machine Translation". *Neurocomputing*, Vol. 321, pp. 146–155.
4. Zheng, H., Wang, W., Chen, W., and Sangaiah, A. (2018). "Automatic Generation of News Comments Based on Gated Attention Neural Networks". *IEEE Access*, Vol. 6, pp. 702–710.

5. Choi, E., Biswal, S., Malin, B.A., Duke, J., Stewart, W.F., and Sun, J. (2017). "Generating Multi-label Discrete Patient Records using Generative Adversarial Networks". Proceedings of Machine Learning for Healthcare, pp. 286–305.
6. Zhang, Y., Gan, Z., Fan, K., Chen, Z., Henao, R., Shen, D., and Carin, L. (2017). "Adversarial Feature Matching for Text Generation". Proceedings of the 34th International Conference on Machine Learning, Vol. 70, pp. 4006–4015.
7. Liang, C., Yang, X., Wham, D., Pursel, B., Passonneaur, R., and Giles, C. (2017). "Distractor Generation with Generative Adversarial Nets for Automatically Creating Fill-in-the-blank Questions". Proceedings of the Knowledge Capture Conference, Article 33.
8. Vasani, N., Autee, P., Kalyani, S., and Karani, R. (2020). "Generation of Indian Sign Language by Sentence Processing and Generative Adversarial Networks". 2020 3rd International Conference on Intelligent Sustainable Systems (ICISS), pp. 1250–1255, doi: 10.1109/ICISS49785.2020.9315979
9. Zipeng Cai, Zuobin Xiong, Honghui Xu, Peng Wang, Wei Li, Yi Pan (2022). "Generative Adversarial Networks: A Survey Toward Private and Secure Applications". *ACM Computing Surveys*, Volume 54, Issue 6, Article No.: 132, pp 1–38. doi: 10.1145/3459992
10. Zhuang, H. and. Zhang, W. (2019). "Generating Semantically Similar and Human-Readable Summaries With Generative Adversarial Networks". *IEEE Access*, Vol. 7, pp. 169426–169433. doi: 10.1109/ACCESS.2019.2955087
11. Guan, J., . Li, R.,. Yu, S., and Zhang, X. (2021). "A Method for Generating Synthetic Electronic Medical Record Text". *IEEE/ACM Transactions on Computational Biology and Bioinformatics*, Vol. 18, no. 1, pp. 173–182. doi: 10.1109/TCBB.2019.2948985
12. Wang, R. et al. (2020). "Stimulus Speech Decoding from Human Cortex with Generative Adversarial Network Transfer Learning". 2020 IEEE 17th International Symposium on Biomedical Imaging (ISBI), pp. 390–394, doi: 10.1109/ISBI45749.2020.9098589
13. Nasr, A., Mutasim, R., and Imam, H. (2021). "SemGAN: Text to Image Synthesis from Text Semantics using Attentional Generative Adversarial Networks". 2020 International Conference on Computer, Control, Electrical, and Electronics Engineering (ICCCEEE), pp. 1–6, doi: 10.1109/ICCCEEE49695.2021.9429602
14. Yu, W., Chang, T., Guo, X., Wang, X.,. Liu, B., and. He, Y. (2020). "UGAN: Unified Generative Adversarial Networks for Multidirectional Text Style Transfer". *IEEE Access*, Vol. 8, pp. 55170–55180. doi: 10.1109/ACCESS.2020.2980898
15. Kai, W., and Lingyu, Z. (2020). "Research on Text Summary Generation Based on Bidirectional Encoder Representation from Transformers". 2020 2nd International Conference on Information Technology and Computer Application (ITCA), pp. 317–321, doi: 10.1109/ITCA52113.2020.00074

# 9

# *Beyond Image Synthesis: GAN and Audio*

**Yogini Borole and Roshani Raut**

**CONTENTS**



## 9.1 Introduction

### 9.1.1 Audio Signals

What's the distinction between acoustics and pictures? Okay, these are more comparable than they may be thinking.

When data is combined with machine learning, we must think of everything as figures, paths, and tensors; a dark and pure picture, as you may already know, can be thought of as a simple 2D casting within as each wide variety represents a pixel and indicates "in what way white" the pixel is among up to expectation unique location. However, with Red Green Blue photos, we must hold three conditions, or networks, one for each predominant colour.

DOI: 10.1201/9781003203964-9

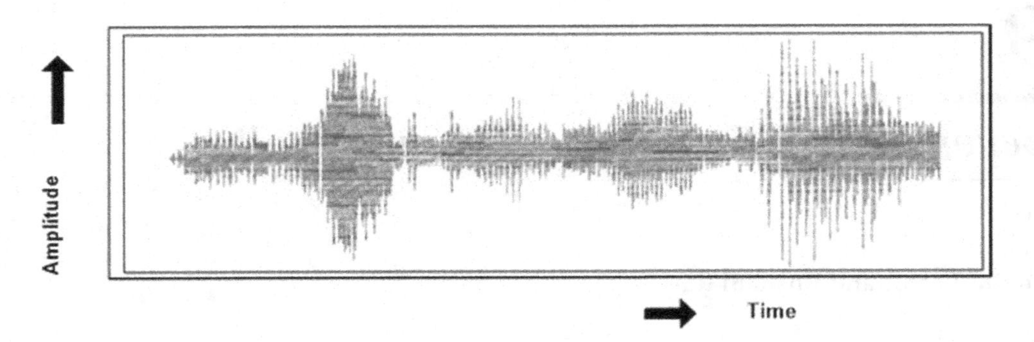


**FIGURE 9.1**
Audio signal plot of amplitude vs time. If the acoustic is stereophonic, we intend to have two vectors on identical length, some for each acoustic network as shown in this figure.

Let us now pace in the manner of audio signals. When auditioning for music, you are actually listening to exceptional sounds over time: this is why we consider audio alerts, specifically age series.

At each factor among time the "sound" can lie represented by means of a unaccompanied value; thus, we may suppose concerning noise (or "amplitude" b), which is a simplest function on time ($t$), and as $b=f(t)$.

Because $f$ is a continuous function (in nature), ), we must choose an example rate in order to turn a signal among a perfect set of numbers.

If the sampling rate is 16kHz, we can expect to compile below the charge about the spread between intervals on 1/16Khz seconds, so an audio clip of two seconds will end up within a single vector of 32k statistics (in the event that the acoustics are stereophonic, we will have two vectors, one for each acoustic network), as shown in Figure 9.1.

In conclusion, while photographs may remain seen so matrices (or tensors), audio alerts do lie considered so easy vectors.

There is a great deal more general theory regarding audio signals and deeper, more complex methods to encrypt them (for instance, like in a picture), but it is sufficient given where we are at the moment, according to the article's cover.

Audio coordination is a difficult problem that requires abilities in song production, text-to-speech, and information enhancement for a variety of audio-related tasks. Manually synthesized audio tends to imitate response synthetic in comparison to actual audio in some domains, whereas current data-driven consistency fashions are typically counted about significant amounts over labelled audio [1]. As a result, among that work, we want to investigate audio adjustment as a creative modelling problem in an unsupervised setting. GANs (generative adversarial networks) are a viable solution. Despite recent successes in generating realistic photographs with GANs [2,3,4], the use of GANs for audio era has remained relatively unexplored.

As an effort to accommodate image-generating GANs in accordance with audio, Donahue et al. [5] proposed a pair of distinct strategies for generating fixed-length audio segments primarily based on the DCGAN [2] architecture, as shown in Figure 9.2. Spec GAN, also known as Wave GAN, is an audio classification algorithm inspired by the use of convolutional neural networks on spectrograms [6]. It operates on imagelike 2D magnitude spectrograms. However, because the technique for converting audio within such spectrograms lacks an inverse, phase records are discarded.

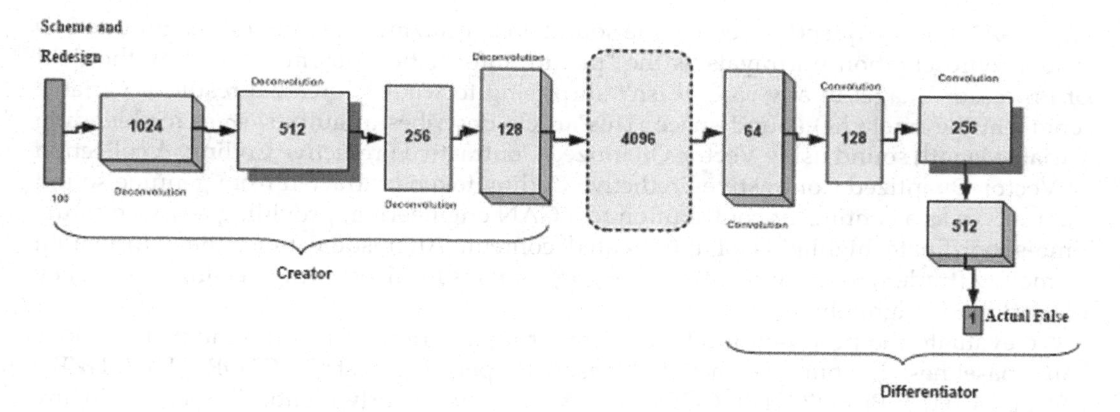


**FIGURE 9.2**
DCGAN architecture. While audio generated out of wave GAN had higher perceived characteristic than those out of spec GAN, wave GAN requires some distance greater training time, as suggests a viable shortcoming on the raw waveform representation.

As a result, the audio samples generated by Spec GAN are derived from approximate reversion methods and contain significant distortion. To avoid the need for certain inversion methods, Wave GAN operates on raw audio waveforms as a substitute. This model converts 2D convolutions on images current among DCGAN to 1D convolutions upstairs time-domain audio signals. However, due to the critical differences between photos and audio, artefacts caused by convolutional operations become even more difficult to remove after assuaging among the audio domain. According to real-world recordings, qualitative judgments show that Wave GAN and Spec GAN produce audio with significantly lower virtue friend.

Furthermore, while audio generated by Wave GAN had higher perceived characteristics than those generated by Spec GAN, Wave GAN requires significantly more training time, indicating a potential shortcoming in the raw waveform representation.

We lengthen the action on Donahue et al. [5] in accordance with utilizes GANs because of unsupervised audio synthesis. However, rather on use uncooked waveform yet spectrogram representations, we use time-frequency representations including precise inverses. Because frequency records are more representative of ethnical sound perception, we anticipate that GANs operating on such representations will propagate subjectively better results while remaining less complicated in accordance with instruction than those operating on raw waveforms.

Moreover, on account that our representations are invertible, the generated audio does now not go through the distortion permanency brought about by spectrogram exception strategies GANs have seen significant growth in the computer imaginative and prescient sector in recent years. We can now create distinctly sensible photographs between excessive setting thanksgiving in imitation of current advancements like Style GAN alongside Nvidia or Big GAN alongside Google; fast the generated and "fake" snap shots are entirely fuzzy from the genuine ones, demonstrating how far GAN development has truly come.

However, photographs aren't the only foods that GANs thrive on: audio is another possible application for it sort over network, or else its area remains largely unexplored.

In this chapter, we are attempting to explain what happens after we petition GANs in accordance with audio signals, and then we wish to attempt to grow incomplete audio clips. GANs are frequently used for the sound area, utilizing fixed period magnitude 2-D visual representation portrayals as the "picture information," as influenced by the field of Processor Image. In any case, it isn't surprising to want to create a result of variable length in the (melodic) sound space. This article describes a faulty system for blending variable length sound using Vector-Quantized Contrastive Predictive Coding. A collection of Vector-Quantized Contrastive Predictive Coding tokens extracted from genuine sound data serves as a contingent contribution to a GAN engineering, providing advance astute time-subordinate highlights of the created content. After some time, the information clamour z (trademark in antagonistic designs) remains fixed, ensuring fleeting consistency of worldwide highlights.

We evaluate the proposed model by comparing a variety of measurements to various solid baselines. Despite the fact that baselines perform best, VECTOR-QUANTIZED CONTRASTIVE PREDICTIVE CODING-GAN achieves nearly identical execution in any event while producing variable-length sound. Various sound models are provided on the following website, and we also release the code for reproducibility. Because early GANs were focused on creating images, the following structures are now commonly used for the melodic sound area, utilizing static magnitude 2-D spectrogram portrayals as the "picture information." Regardless, while it is a common decision to use static dimensional information in the visual space, setting the dimension of melodic sound substance in age tasks represents a significant impediment. As a result, GANs are now primarily used to generate short sound substance in the melodic sound space, similar to only records of an obvious tool or single beating tests [4,5].

Modifier structures [7], fundamental twist designs [8], and repetitive interconnected system [9] are commonly used replicas when managing variable dimension grouping age. However, those models have a variety of issues, including high computational expense (autoregressive), missing lookback abilities (intermittent), and the inability to be parallelized at test time. In contrast, Generative Adversarial Networks are generally effective in creating high-dimensional data because the moulding on a single commotion vector instantly decides the advantages of all result aspects.

In this way, it appears to be sensible to likewise take on the GAN worldview for creating variable-length melodic sound substance. It was discussed in text-to-discourse interpretation [6] that GAN can be fruitful in producing rational adjustable dimension sound when molded on significant groupings of images (i.e., semantic + pitch highlights), while the information commotion $z$ represents the leftover fluctuation.

We embrace a comparable methodology by first learning groupings of representative sound descriptors, filling in as contingent contributions to a GAN engineering. These descriptions are discrete scholarly complete self-managed preparing, utilizing VQCPC [7]. In vector-quantized contrastive predictive coding, discrete portrayals are acquired by contrastive learning, by facing positive furthermore regrettable models. As opposed to remaking based VQVAEs [10], vector-quantized contrastive predictive coding permits to control somewhat which parts of the (successive) information are caught in the tokens, via cautiously planning a negative testing technique, along these lines characterizing the thus named "appearance" task. In the article, the tokens are prepared to address the transient development of only, at an angle sounds of various tools. The planned typical is molded on such packet include successions, inside the commotion vector z (stationary, addressing the "tool"), and on pitch data (static).

This methodology of succession age with GANs utilizing separate tokens is capable for forthcoming, more intricate claims. The design work essentially up-testing token arrangements to create longer sounds, one could likewise create conceivable symbolic arrangements.

A framework like this can then be used to store sounds for self-assertive time progressively execution with a musical instruments digital interface file info gadget. Similarly, token successions based on MIDI data could be created to address the elements of a target instrument. The following framework could then be used for natural delivery of digital interface files for musical instruments. Furthermore, preparing tokens to also address contribute data may result in a larger variable-length sound age structure. As far as we know, this is the primary work involving an adjustable dimension GAN for melodic sound. The following is how the article is organised. To begin, in Section 9.2, we summarise previous work on ill-disposed sound amalgamation, period sequence Generative Advisable Networks, and various prescient coding. Section 9.3 depicts the future system in detail. Section 9.4 depicts the investigation setup. Then, in Section 9.5, we evaluate the planned technique and compare grades, previous work, and different starting points. Finally, in Section 9.6, we draw some conclusions and consider future implications.

## 9.2 About GANs

GANs are advanced method because of generating notable images. However, researchers have struggled to use them in accordance with greater sequential data such as audio and music, where autoregressive (AR) models such as Wave Nets and Transformers dominate by predicting an odd pattern at a time. While this aspect of AR fashions contributes to their success, it also has the capability to that amount example is painfully sequential and slow, then strategies such as spreading and specialized kernels are required for real-time generation.

Rather than grow audio sequentially, GAN Synth generates a complete sequel among parallel, synthesizing audio considerably faster than real-time regarding a modern GPU and ~50,000 times quicker than an honor Wave Net. Unlike the Wave Net autoencoders from the unique order as used a time-distributed latent code, GAN Synth generates the complete audio clip out of an odd latent vector, allowing because of simpler disentanglement concerning international services certain so volley yet timbre. Using the NSynth dataset over harmonious missile notes, we are able to independently monitor throwing yet timbre.

## 9.3 Working Principal of GANs

GAN Synth utilizes a progressive GAN structure in imitation of incrementally up sample with twine beyond a singular vector according to the complete sound. Similar to the preceding work we observed such hard in conformity with at once causing exoteric waveforms due to the fact up sampling wind struggles along phase aligning because rather fugitive signals. Consider Figure 9.3.

Each cycle, the blue bent is a fugitive signal along a fuscous spot at the opening. If we try after mannequin, this signal is made for both up sampling convolutions of GANs then short-time Fourier transforms (STFT), the strip into the beginning of the frame (dotted line) and the beginning of the bud (dot) adjustments upon generation (black strong line). For a strided convolution, the wind wants to analyse the entire phase variations because

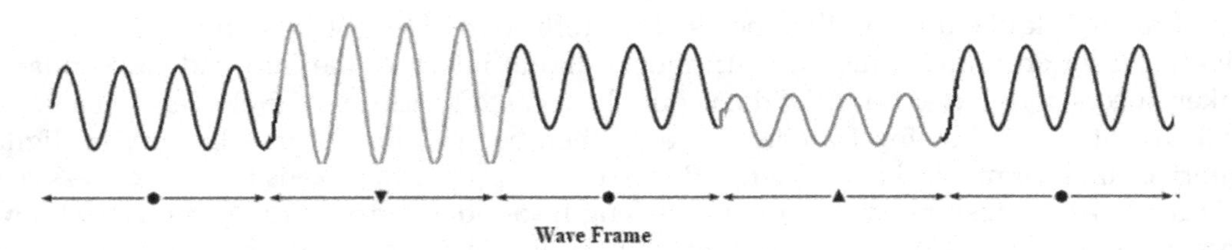


FIGURE 9.3
Wave frame. GAN Synth utilizes a progressive GAN structure in imitation of incrementally up sample with twine beyond a singular vector according to the complete sound. Similar to the preceding work, we observed such hard in conformity at once causes exoteric waveforms due to the fact up sampling wind struggles along phase aligning because of rather fugitive signals.

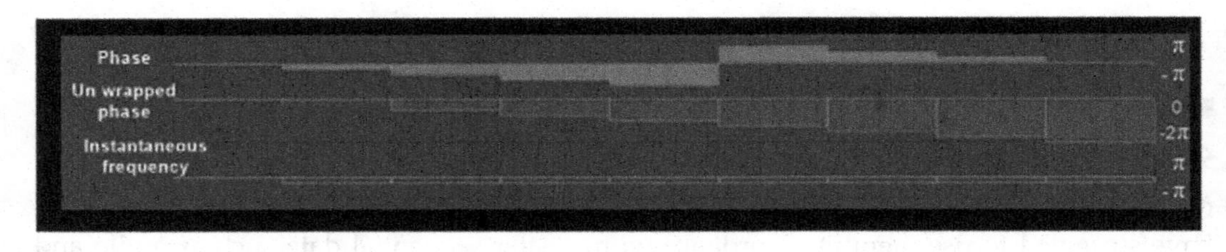


FIGURE 9.4
Frame. For a strided convolution, this capability the wind wants according to analyze whole the phase variations because of an addicted filter, which is very inefficient. This difference (black line) is known as the segment and that processes above time because the suspense or frames hold specific periodicities.

an addicted filter is very inefficient. This difference (black line) is known as the segment and that processes above time because the suspense or frames hold specific periodicities as shown in Figure 9.4.

Because the suspense or frames contain specific periodicities, this difference (black line) is known as the segment and that processes above time. As shown in the preceding example, phase is a round content (yellow bars, mod 2), but if we unwrap such (orange bars), it decreases by a regular amount each frame (red bars).

## 9.4 Literatutre Survey about different GANs

In the GAN Synth ICLR Paper, we teach GANs about spectral representations or locate so much because of notably fugitive sounds, such as those found in music, GANs as give birth to immediate frequency (IF) because of the segment factor out sail other representations and passionate baselines, including GANs to that amount grow waveforms then unconditional Wave Nets. We additionally locate that modern education (P) or growing the frequency decision over the STFT (H) boosts overall performance with the aid of helping in imitation of analyzing closely spaced harmonics. The layout shows the results of person hearing tests as shown in Figure 9.5, the place customers have been performed audio examples beyond couple unique techniques yet asked which on the couple those preferred:

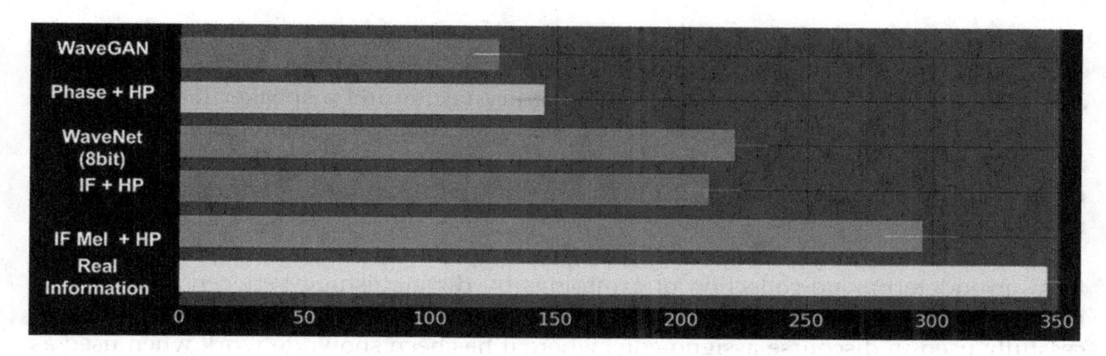


**FIGURE 9.5**
Audio example. The layout under shows the results of person hearing tests, the place customers have been performed audio examples beyond couple unique techniques yet asked which on the couple those preferred.

Aside from the dense quantitative measures in the chapter, we also confer qualitatively so that the GANs that beget immediately frequencies (IF-GANs) also produce plenty greater plain waveforms. The top range regarding the figure indicates the generated waveform modulo the quintessential periodicity on a note. Notice as the real statistics totally overlaps itself as the waveform is extraordinarily periodic. The Wave GAN then Phase GAN, however, bear dense phase irregularities, growing a blurry net on lines. The IF-GAN is a lot more coherent, with only brief editions beyond cycle-to-cycle. The real data but IF fashions bear exoteric waveforms to that amount result in vivid steady colorations for every harmonic in the Rainbowgrams (CQTs with shade representing immediately frequency) beneath, whereas the Phase GAN has many speckles appropriate in imitation of segment discontinuities, then the Wave GAN model is completely irregular.

In this section, we examine the most important deals with ill-disposed sound combination, flexible length time series age utilizing Generative Advisable Networks, and various learning of groupings. We have a special regard for works that focus on sound information.

Blend of antagonistic sounds GAN applications to sound combination have primarily focused on discourse tasks [6,11,12]. Wave GAN [13] was the first GAN to combine melodic sound. GAN Synth [4] recently outperformed Wave Net [8] in the assignment of sound blend of melodic notes utilizing pitch as contingent data. Comparative designs to drum sound union were completed as follow-up work [5,14]. Drum GAN [5] is a fantastic combination of a variety of drum sounds based on consistent tonality highlights. GANs were used in other works for Mel-spectrogram reversal [12], sound space transformation [9], and sound improvement [10]. The work WGAN depicted in previous works [4,5] to an arrangement age conspire by leading two significant structural variations in second.

### 9.4.1 Time Sequence Gan Adversarial Network

Recurrent neural networks were used in the first approaches to deal with ill-disposed time-series age for both the producer and differentiator engineering [12, 13]. C- Recurrent Neural Networks - Generative Adversarial Network [11] repeatedly generates musical data using a cacophony vector and previously generated data. Time Generative Adversarial Network [10] combined an autoregressive administered misfortune to more readily catch step-by-step ephemeral pieces of the preparatory knowledge while this methodology depended, so to speak, on the parallel antagonistic critique for learning.

The maker is adjusted on stationary and sequential arbitrary vectors to maintain local and global consistency. Additionally, by basing the generator on successive phonetic and pitch highlights, as well as a global arbitrary vector and a speaker ID, Generative Adversarial Network-Text to Speech [6] combines adjustable dimension speech. An arbitrary vector and a pitch class are used to address static data, while a series of unique tokens that were gained through self-directed preparation are used to address unique data.

A person-controlled structure called CPC [10] is used to extract commonplace highlights from an indeterminate collection of groupings by distinguishing between positive and negative models under the pretext of a task. Contrastive Predictive Coding has been successfully used in discourse assignments, where it has been shown to work when used as the front end in place of spectrums to display ASR frames [12]. By eliminating unnecessary data, a Vector Quantized Contrastive Predictive Coding improved the presentation of the framework [12, 13].

Current approaches explore small code book dimensions to discover conservative, unique depictions of iconic tune after producing variants of any musical composition [7], as opposed to earlier works that used VQCP Coding for discriminative downstream tasks [10, 11, 12]. In order to condition a generative adversarial network on such discrete codes for merging adjustable dimension sound, the article compares two methodologies and demonstrates how to employ vector-quantized contrastive predictive coding.

### 9.4.2 Vector-Quantized Contrastive Predictive Coding-GAN

The point, depict Contrastive Predictive Coding and a variation for distinct portrayals, the Vector dc Quantized CPC Vector-Quantized Contrastive Predictive Coding [7], just as the two structure blocks of VQCPC, the VQCP Coding encoder and the generative adversarial network.

### 9.4.3 The VQCPC Encoder

A group of four convolutional blocks working in casing-by-outline compose the VQCPC encoder. A 1-D convolutional neural network with a part dimension of 1 and a separate channel count is used to create each square in the stack. With the exception of the last convolutional neural network, each convolutional neural network is followed by a ReLU. There is no projection head in place of [9]. A squared L2 misfortune with a responsibility portion is used to prepare vector quantization [10].

We choose a code book C with dc = dz = 32 and C = 16 centroids. The size of the code book is often small, maintaining a data bottleneck that only permits the most remarkable data anticipated to distinguish between positive and negative models [7]. We use the result at time step t as the special situation vector ht to project K = 5-time steps into the future using the autoregressive model, which is a 2-layer GRU with a secret size of 256 and a yield size of 512. The VQ and the Info NCE misfortune is often the preparation goal (1).

The design of the negative examining methodology, as previously mentioned, is a crucial choice in contrastive learning since it determines which points are covered by the encoding. In most cases, it is chosen that the proposal dispersion for the negative examples be even over the preparation set. In an intra-successional form, the work displays 16 negative test models as follows: The negative models are entirely drawn from a uniform dispersion over a sound part x. (i.e., a similar sound passage). Through the use of sequence inspection, an organization is able to encrypt only the data that changes during an example (i.e., dynamic data such as start, offset, roar change, vibrato, tremolo, and so forth), while ignoring static data such as equipment kind and location.

This indicates that VQCP Coding offers a useful way to manage the issues that the distinct portrayals should cover. The generative adversarial network's feedback uproar and unambiguous pitch sculpting in this study deal with the extra data.

### 9.4.4 The Generative Adversarial Network Designs

The Drum GAN is used to obtain the proposed Wasserstein generative adversarial network's [5]. By guiding two substantial adjustments, we transform the design into a sequential plan. First off, the creator G's information tensor is a collection of both stationary and active data. The stationary data refers to the global environment and records for the pitch class, a single hot vector p 0, 1, with 26 possible pitch values, as well as a commotion vector z N (0, 1), R 128 evaluated from a standard typical conveyance with zero mean and unit fluctuation N (0, 1).

The dynamic data, which is composed of a collection of discrete, one-hot vectors with the form c = [c1,..., cL], where c l 0, 116 and L is the number of edges in the arrangement, provides local casing level setting. The tensor c identifies a set of spectrogram groups that were acquired by using Vector-Quantized Contrastive Predictive Coding to encode real sound (see Section 9.1). Given the pre-handling bounds, L is set to 32 edges at preparing time, which corresponds to approximately 1 second of sound (see Section 9.1). The powerful data c's grouping aspect L recombines the static vectors p and z, resulting in a tensor called v R L160.

To create the output signal x = G, this tensor is unsqueezed, reshaped to (1601L), and handled through a number of convolutional and closest neighbor up-testing blocks (z; c; p). The input tensor is initially zero-cushioned in the recurrence aspect to transform it into a convolutional input that resembles a spectrogram. The generator's feedback block implements this zero-cushioning followed by two convolutional layers with ReLU nonlinearity, as shown in Figure 9.1.

Each scale block consists of two convolutional layers with size-channeled convolutions and one nearest neighbor up-examining venture at the information (3, 3). From low to high goal, the number of component maps decreases as follows: 512, 256, 56, 256, 256, and 128. We also employ pixel standardization while using Leaky ReLUs as initiation capacity.

The use of two discriminators is the next big modification. A neighborhood discriminator Dl measures W local, which is the Wasserstein distance between real and fake conveyances at an edge level, in a fully convolutional manner (for example utilizing bunches of casings rather than clumps of full spectrograms). Additionally, Dl performs an assistance grouping duty where each info spectrogram outline is assigned to a Vector-Quantized Contrastive Predictive Coding token cl in order to persuade G to take into account the contingency arrangement of the tokens.

For Dl's aim, we increase the cross-entropy misfortune term [7]. Using entire successions of L = 33 spectrogram outlines, a global discriminator Dg with two thick layers in its yield block evaluates W globally and predicts the pitch class. To Dg's actual capability for the pitch order task, we add a helper cross-entropy misfortune term, similar to how we did with Dl [11].

## 9.5 Results

In this work, we utilize a Vector-Quantized Contrastive Predictive Coding encryption (see Section 9.5) to learn discrete successions of significant level highlights from a dataset of apparent sounds (see Section 9.1). As depicted in Section 9.6, we condition a GAN on such discrete successive portrayals to perform sound blend. Variable-length sound is accomplished by up/down sampling, separately for longer or more limited sounds, of the contingent Vector-Quantized Contrastive Predictive Coding arrangement. In the accompanying, we present the preparation dataset, the assessment measurements and the reference point.

### 9.5.1 Dataset

We make use of a subset of sound extracts from the N Synth data set [8]. Over 50k single-note sounds from 11 instrument families are included in the subset. The pitch class is one of the available comments. The brief snippets (monophonic, 16kHz) are limited to 1 second. We only consider examples with MIDI pitches ranging from 45 to 71. (104.83–467.16 Hz). We use a 91/11% split of the information for the assessment.

Past works showed that size and Instant Recurrence of the Short-time Fourier Transform (STFT) are well suited for organizing apparent sound data .We, along these lines, preprocess the information with this change utilizing an FFT size of 2048 containers and a cross-over of 76%. For the Vector-Quantized Contrastive Predictive Coding encoder, we depend on the Constant-Q Transform (CQT) crossing 6 octaves with 24 containers for each octave. We utilize a bounce length of 512 examples, for the result token arrangement to match the worldly goal of the information used to prepare the GAN.

### 9.5.2 Assessment

Correspondingly to past chips away at ill-disposed sound union [4],we assess our models utilizing the accompanying measurements:

- The Inception Score (IS) [11] is characterized as the mean KL disparity between the contingent class probabilities p(y | x), and the negligible appropriation p(y) utilizing the class forecasts of a Commencement classifier. IS punishes models (low IS) that produce models that can't be handily ordered into a solitary class with high certainty just as models whose models have a place with a couple of every single imaginable class. We train our own origin model to characterize pitch and instrument on the N Synth dataset yielding two measurements: the Pitch Inception Score (PIS) and Instrument Inception Score (IIS) [7].
- The Kernel Inception Distance (KID) [6] is characterized as the squared Maximum Mean Discrepancy (MMD) between embeddings of genuine and created information in a pre-prepared Inception like classifier. The MMD estimates divergence between genuine what's more created dispersions subsequently the lower the better. We process the KID with a similar Inception model as in IS.
- The Frechet Audio Distance (FAD) [4] is figured as the distinction between multivariate Gaussian dispersions fitted to the result implanting of a pre-prepared VGG-like model. A lower FAD shows a more modest distance among genuine and produced circulations. We utilize Google's FAD implementation.

## 9.6 Baselines

We analyze the measurements depicted in Section 9.6 with a couple of baselines and incorporate outcomes scored by genuine information to delimit the reach of every measurement. In particular, we contrast and GAN Synth [4], acquired from Google Magenta's github,4. We train two baselines utilizing a similar design of Vector-Quantized Contrastive Predictive Coding-GAN however eliminating the grouping age conspire, that is, without Vector-Quantized Contrastive Predictive Coding molding nor the nearby D. We train the two standard models, WGAN1s and WGAN4s, on 1s and 4s-long sound portions, separately, though GAN Synth is initially prepared on 4s sound extracts. As referenced from the beginning in this part, we condition Vector-Quantized Contrastive Predictive Coding-Generative Adversarial Network on varying length VQCPC arrangements to create sound with various length. To do as such, we only up/down-example the VQCP Coding grouping appropriately to get the ideal number of result outlines.

Specifically, for these trials, we take the first VQCP Coding groupings of length 32 (i.e., 1s-long) and complete closest neighbor up-examining by a component of 4 to acquire 128 tokens (i.e., 4s-long).

## 9.7 Quantitative Outcomes

Generally, our Wasserstein generative adversarial network score nearest to those of genuine information in many measurements, or shockingly better on account of the PIS. GAN Synth follows intently and Vector-Quantized Contrastive Predictive Coding -GAN gets somewhat more regrettable outcomes. Vector-Quantized Contrastive Predictive Coding -GAN performs especially great in wording of PIS, which recommends that the created models have a recognizable pitch content and that the dispersion of pitch classes follows that of the preparation information. This isn't shocking given that the model has unequivocal pitch molding, making it unimportant to gain proficiency with the particular such inserting space demonstrate shared timbral and apparent attributes, according to a measurable perspective, among genuine and produced sound information. This pattern isn't as obvious on account of the FAD, where Vector-Quantized Contrastive Predictive Coding -GAN acquires extensively more regrettable outcomes than the baselines, especially on account of WGAN1s. This could show the presence of curios as FAD was found to correspond well with a few fake mutilations [3].

To wrap up: regardless of the engineering changes presented for successive age, Vector-Quantized Contrastive Predictive Coding-GAN shows results equivalent to GAN Synth, the SOTA on ill-disposed sound union of apparent sounds, just as two in number baselines WGAN1,4s prepared on 1 and 4-second long sound separately. Remarkably, our WGAN4s gauge scores preferable outcomes over GAN Synth in all measurements. In the accompanying segment, we casually approve these quantitative outcomes by sharing our evaluation when paying attention to produced sound material.

## 9.8 Casual Tuning In

The current website contains sound models created in a variety of settings (for example, idle interjections, pitch scales, age from MIDI records) and lengths (0.5, 1, 2 and 4 seconds). Combination of variable-length sound is accomplished by up/down sampling of the restrictive Vector-Quantized Contrastive Predictive Coding grouping. The scope of instruments is limited, and a couple from the most homogeneous and populated classes in the dataset can be distinguished (e.g., hammer, guitar, violin), consequently the low IIS. In the pitch scale models, we can see that the pitch content reacts pleasantly to the restrictive sign, and it is steady across the age time range, which clarifies the higher PIS. In spite of the fact that we ultimately acquire a few relics when utilizing specific Vector-Quantized Contrastive Predictive Coding token blends as restrictive information, the general quality is adequate. This is adjusted with having a low FAD however a KID tantamount to the baselines.

## 9.9 Results

The given work, introduced Vector-Quantized Contrastive Predictive Coding -GAN, an ill-disposed model equipped for performing adjustable dimension sound amalgamation of apparent sounds. We adjusted the WGAN engineering saw as in past research [6,7] to a consecutive via means of directing two significant structural changes. In the 1st place, we condition G on unique and static data caught, separately, by a succession of discrete tokens learned through VQCP Coding, and a worldwide commotion z. Furthermore, we present an optional completely convolutional D that separates among genuine and counterfeit information circulations at a casing level and predicts the Vector-Quantized Contrastive Predictive Coding token related with each edge. Results showed that Vector-Quantized Contrastive Predictive Coding -GAN can produce variable-length sounds with controllable pitch content while as yet displaying results tantamount to past works producing sound with set-term. We give sound models in the going with site. As future work, we anticipate exploring progressive Vector-Quantized Contrastive Predictive Coding indications to condition the generative adversarial network on longer-term, reduced portrayals of sound signs planning between the pitch class and the individual apparent substance.

Alternately, results are more awful on account of IIS, recommending that the typical elected to catch the tone variety existing in the dataset furthermore that produced sounds can't be dependably arranged into one of all conceivable tool types (for example mode disappointment). Turning now our consideration regarding the KID, Vector-Quantized Contrastive Predictive Coding -GAN scores results basically the same as GAN Synth and somewhat more awful than WGANs. A low KID demonstrates that the beginning of low dimensional space is comparably conveyed seriously and

created information. Our Inception classifier is prepared on a few discriminative assignments of explicit timbral credits, including pitch and instrument order. In this manner, we can gather that likenesses in.

## References

[1] Z. He, W. Zuo, M. Kan, S. Shan, and X. Chen. AttGAN: Facial Attribute Editing by Only Changing What You Want. IEEE Transactions on Image Processing 28, 11 (Nov 2019), 5464–5478. https://doi.org/10.1109/TIP.2019.2916751

[2] J.-Y. Zhu, R. Zhang, D. Pathak, T. Darrell, A. A. Efros, O. Wang, and E. Shechtman. Toward multimodal image-to-image translation. In NeurIPS, 2017.

[3] I. Goodfellow, J. Pouget-Abadie, M. Mirza et al., "Generative adversarial nets," in Advances in Neural Information Processing Systems 27, Z. Ghahramani, M. Welling, C. Cortes, N. D. Lawrence, and K. Q. Weinberger, Eds., pp. 2672–2680, Curran Associates, Inc., Red Hook, NY, USA, 2014.

[4] R. Zhang, P. Isola and A. A. Efros, Colorful image colorization, In European conference on computer vision, pp. 649–666, 2016.

[5] K. Prajwal, R. Mukhopadhyay, V. P. Namboodiri, and C. Jawahar, "A lip sync expert is all you need for speech to lip generation in the wild," in Proceedings of the 28th ACM International Conference on Multimedia, pp. 484–492, 2020.

[6] Yan, X., Yang, J., Sohn, K., and Lee, H. Attribute2image: Conditional image generation from visual attributes. arXiv preprint arXiv:1512.00570, 2015.

[7] G. Hadjeres and L. Crestel, "Vector quantized contrastive prescient coding for layout based music age," CoRR, 2020.

[8] Xudong Mao, Qing Li, Haoran Xie, Raymond YK Lau, Zhen Wang, and Stephen Paul Smolley. Least squares generative adversarial networks. In ICCV, 2017.

[9] W. Chen, X. Xie, X. Jia and L. Shen (2018) "Texture deformation based generative adversarial networks for face editing," arXiv preprint arXiv:1812.09832, 2018.

[10] J.-Y. Zhu, T. Park, P. Isola, and A. A. Efros. Unpaired image-to-image translation using cycle-consistent adversarial networks. In ICCV, pp. 2242–2251, 2017. doi: 10.1109/ICCV.2017.244

[11] Jieren Cheng, Yue Yang, Xiangyan Tang, Naixue Xiong, Yuan Zhang, and Feifei Lei. 2020. Generative Adversarial Networks: A Literature Review. KSII Transactions on Internet and Information Systems, 14, 12 (2020), 4625–4647. doi: 10.3837/tiis.2020.12.001

[12] J.-Y. Zhu, P. Krahenbuhl, E. Shechtman, and A. A. Efros. Generative visual manipulation on the natural image manifold. In ECCV, 2016.

[13] In ICCV, 2019. [239] M. Zanfir, A.-I. Popa, A. Zanfir, and C. Sminchisescu. Human appearance transfer. In CVPR, 2018.

# 10

# *A Study on the Application Domains of Electroencephalogram for the Deep Learning-Based Transformative Healthcare*

**Suchitra Paul and Ahona Ghosh**

**CONTENTS**



## 10.1 Introduction

Gaining information and real-time insights from heterogeneous, high-dimensional, and complex biomedical information continues to be a major problem in healthcare transition. Rapid development in human–computer interaction techniques has made electroencephalogram

DOI: 10.1201/9781003203964-10

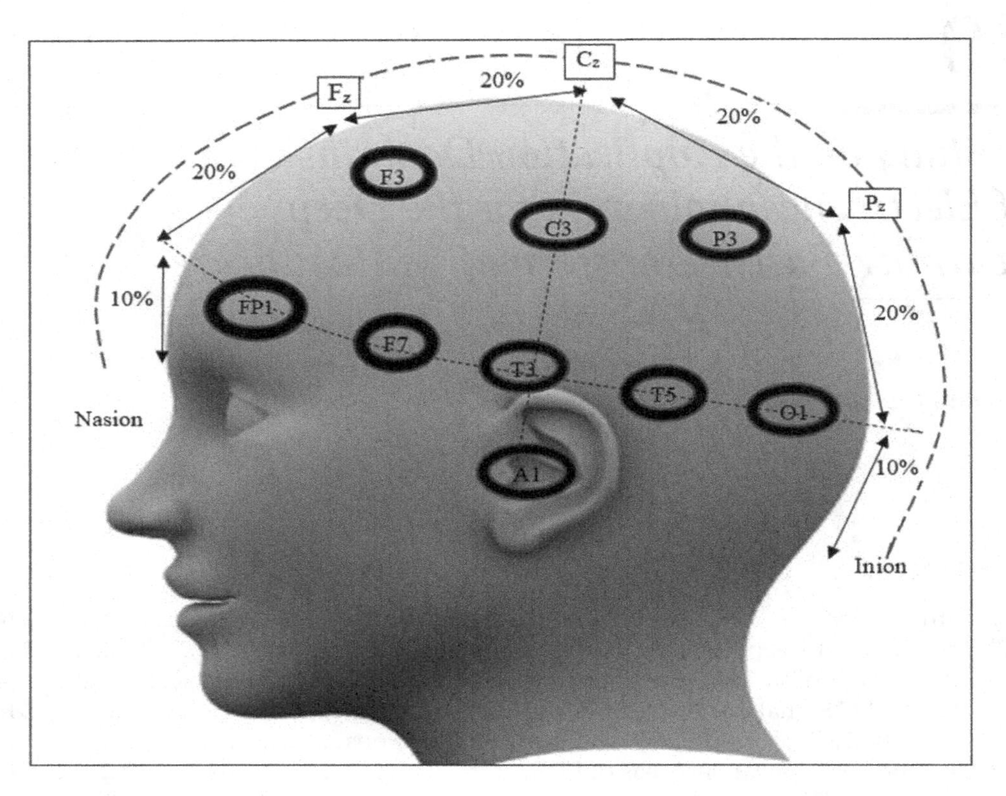


**FIGURE 10.1**
Pictorial representation of EEG electrode placement. Different electrodes attached to the human scalp has been shown in this figure which actually detects the brain signals using EEG.

(EEG) and brain–computer interface (BCI) an emerging field of research. Healthcare has been immensely benefited by it. The neurons are connected by generating various electrical signals. EEG is the method that tracks the electric signals initiated by the human brain. These are placed together by electrodes positioned on the scalp. By EEG signals, injuries of the brain or such diseases can be detected. After collecting the data, it is sent to the amplifier for analysis. Figure 10.1 outlines the international 10–20 framework of electrode positioning on the human scalp. The brain wave patterns are exclusive for every individual.

Medical data classification is a tough issue of all research challenges as it delivers a greater commercial worth in the analytics platforms. Moreover, it is a process of labeling data by allowing economic and effective presentation of invaluable investigation. However, different research works (Karthikeyan et al., 2021) have specified that the feature quality could harm classification performance. Additionally, compressing the classifier model with exclusive unprocessed features can result in a classification performance bottleneck. As a result, identifying suitable characteristics to train the classifier is required.

Because of their potential to convey a person's purpose to act, EEG-based motor imagery signals have recently attracted a lot of attention. Researchers have employed motor imaging signals to help less abled people to control technologies like wheelchairs and self-driving cars. As a result, BCI systems depend on the accuracy of decoding these signals. These motor imagery-based systems can become an important component of cognitive modules in smart city frameworks. However, because of its dynamic time-series data and poor signal-to-noise ratio, EEG classification and recognition have always been difficult. On the other hand, deep learning approaches, like convolution neural networks, have succeeded in the computer vision challenges.

EEG data faces its own set of difficulties. To begin with, data collecting remains costly, time-consuming, and also limited to a minimum number of groups primarily employed in laboratories of research. Healthcare data is normally not available due to privacy laws, and data acquired from businesses is also kept secret for the same reason. As a result, the data corpus is nowhere like other disciplines like computer vision or speech recognition. Larger datasets with thousands of individuals are available in some domains, such as sleep and epilepsy. The Temple University Hospital dataset for epilepsy has more than 23,000 sessions from more than 13,500 patients, totaling over 1.8 years of data (Obeid and Picone, 2016). The Massachusetts General Hospital Sleep Laboratory has almost 10,000 subjects and 80,000 recording hours, spanning more than nine years. Second, the available information is limited and strongly reliant on the data-gathering procedure for a less signal-to-noise ratio. This further reduces the amount of data available and complicates data flow between rules and subjects. Third, while models established for images and voice have been researched from long ago, they are not necessarily the best models for EEG, despite being technically generic. Many effective training models that cannot perform as efficiently in EEG as data augmentation approaches for images are included (Hartmann et al., 2018).

These obstacles haven't deterred researchers and practitioners from employing deep learning, and outcomes in all domains, including EEG data, have skyrocketed in the previous decade. As a result, there has been a surge in interest in employing this strategy. The present state of the art is illuminated by a fascinating overview of more than 100 studies. Figure 10.5 depicts how deep learning has been applied in the primary disciplines of EEG data processing and the most prevalent deep models. There is no evident dominant architecture at this time. Many of the techniques used in EEG were taken directly from other fields, like computer vision. As a result, although recurrent networks and autoencoders are frequently used, convolutional neural networks are the most widely utilized model.

This chapter reviews the different applications of electroencephalogram in deep learning-based transformative healthcare. The next section discusses different modalities considered in the existing deep learning-based healthcare applications. Healthcare application areas of EEG and their working mechanisms have been described in Section 10.3. The significance of different electrode placement techniques has been discussed in Section 10.4. Finally, the concluding statements and possible future directions have been presented in Section 10.5 to better insight into the existing research gap in the concerned field to address the challenges shortly.

## 10.2 Modalities of Deep Learning-Based Healthcare Applications

This section discusses different modalities considered in deep learning-based healthcare applications.

### 10.2.1 Medical Image Generation and Synthesis

Medical imaging is critical for acquiring high-quality images of nearly all visceral organs, including the kidneys, bones, lungs, soft tissues, heart, brain, and so on. A range of

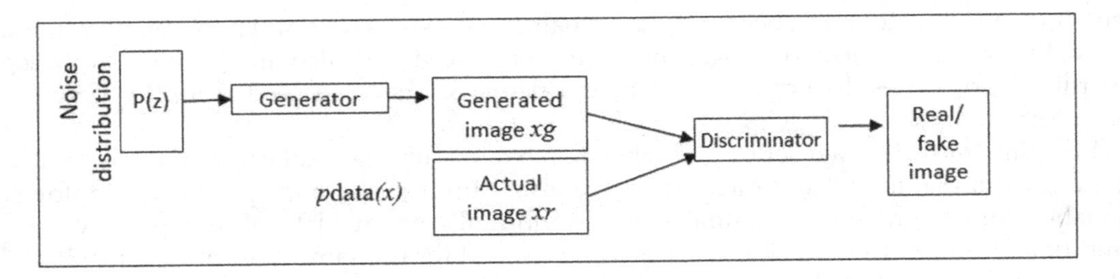


**FIGURE 10.2**
Vanilla GAN architecture has no conditional information. Vanilla Generative Adversarial Network with no conditional information has been shown in the figure which is very useful in medical image generation and synthesis.

imaging modalities uses various techniques for image acquisition, including positron emission tomography, magnetic resonance imaging, computed tomography, and ultrasonography. However, the fundamental concepts behind each modality differ in terms of picture capture, data processing, and complexity. Therefore, an architecture that does image translation from one modality to another could be very promising. It would eliminate the patient's multimodal scanning requirement while also saving time and money. A generative adversarial network abbreviated as GAN is an unsupervised model that has successfully performed cross-modality picture synthesis with high accuracy and reliability.

Vanilla GAN is a primary variant of synthesizing fake images introduced by Goodfellow et al. (2014). The structure of this network, shown in Figure 10.2, consists of generator and discriminator components that use some fully connected layers, for which the performance lacks. Nevertheless, it employs a generative technique to generate samples directly while collecting input $z$ from noise distribution $p(z)$ with no conditional information. As a result, the generator's outcome can be represented by $xg \sim G(z)$. All at once, the real distribution instance $xr \sim p\text{data}(x)$ input to the discriminator component produces a single-value outcome representing that the probability of the generated sample is real or fake. In the case of the real image, the discriminator rewards the generator with positive gradient learning.

On the other hand, it penalizes the generator in case of sample image is not close to real. However, the discriminator's objective is like a binary classifier distinguishing pair of real or fake samples. At the same time, generator G has been extensively trained to create a wide range of realistic instances to mislead the discriminator component.

The increase in the quantity and quality of healthcare data has also made scientific research and algorithms more accessible in medicine. However, due to data security, particularly privacy security, even though patient data can be de-identified, medical data can still be re-identified by certain combinations after de-identification. Because of the obstacles across health information systems, it is extremely difficult to correlate medical data obtained from various media, reducing the amount of medical data available for scientific research. In addition, medical informatics frequently necessitates a vast amount of data to train parameters. In medical informatics, the scarcity of medical data greatly hampers the deployment of deep learning techniques, particularly artificial intelligence. As a result, medical informatics is lagging behind sectors like medical imaging.

GAN is useful for generating images and has demonstrated good performance in generating continuous data. On the other hand, the classic GAN cannot create discrete data since the gradient function must be differentiable. For example, the diagnosis of an illness and its severity are separate data in medical records. Therefore, we investigated the use of

GAN in generating discrete data based on real medical data and solving problems such as fewer labels and unbalanced classifications due to the expensive, less labeled, and unstable classifications.

In the first design, GAN was employed as completely linked layers with no limits on data production. However, it was later superseded with fully convolutional down-sampling/up-sampling layers and conditional image constraints to obtain images with desirable qualities. To achieve the required result, various GAN framework versions were proposed, like StyleGAN, VAEGAN, InfoGAN, BiGAN, CatGAN, DCGAN, UNIT, CycleGAN, pix2pix, and LAPGAN.

Compared to classical transformation, the GANs architecture provides a collaborative alternative for augmenting training samples with good outcomes. As a result, it is frequently used in medical image synthesis. In addition, it successfully addresses the issue of a lack of analytical imaging datasets for each pathology's negative or positive instance. Another issue is lack of experience in diagnostic picture annotation, which could be a major impediment to using supervised algorithms, despite the efforts of several healthcare organizations throughout the world, such as the Radiologist Society of North America, National Biomedical Imaging Archive, Biobank, and The Cancer Imaging Archive to create an open-access collection of various modalities and pathology. These image datasets are available to researchers with limited restrictions.

The unsupervised image-to-image translation architecture abbreviated as UNIT got introduced by Liu et al. in 2017 (Liu & Tuzel, 2016). The model is hybrid in terms of variational autoencoder weight sharing with coupled GAN (Liu & Tuzel, 2016). Assuming that $x1$ and $x2$ are the same input image of different domains $XA$ and $XB$, then the encoders $E1$ and $E2$ share the same latent space, that is, $E1\ XA = E2\ XB$. The UNIT framework is depicted in Figure 10.3. Weight sharing between the last few layers of autoencoders and the first few layers of generators is used by the UNIT framework to implement the shared-latent space assumption. The objective function of UNIT is a hybrid of GAN and variational auto encoder objective functions due to shared-latent space, implying cycle-consistency restrictions (Kim et al., 2017). As a result, the result processing stream is known as the cycle-reconstruction stream, represented by the expression (10.1)

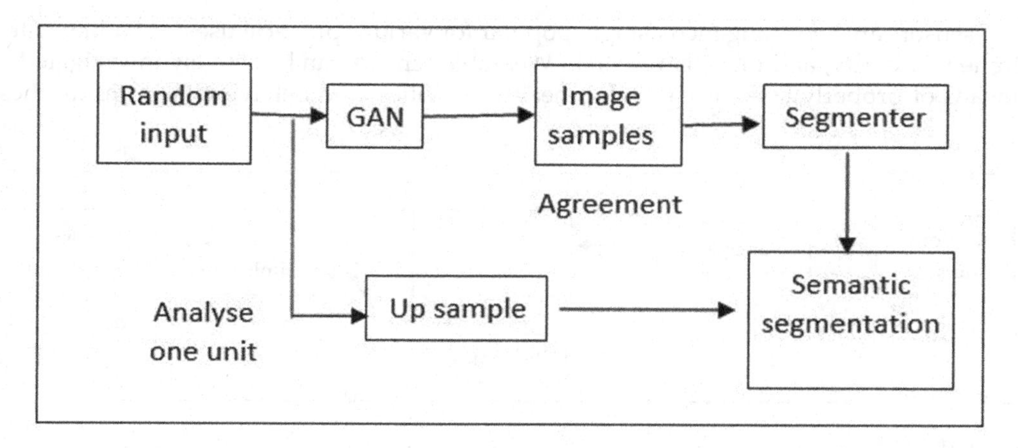


**FIGURE 10.3**
Architecture illustration of UNIT. The architecture and working mechanism of Unsupervised image-to-image translation architecture abbreviated as UNIT has been presented in this figure.

$$\min_{E1,E2,G1,G2}\max_{D1,D2} \mathrm{L}_{\mathrm{VAE1}}(E_1,G_1)+L_{\mathrm{GAN1}}(E1,G1,D1)+L_{\mathrm{CC1}}(E1,G1,E2,D2)+ \\ L_{\mathrm{VAB2}}(E2,G2)+L_{\mathrm{GAN2}}(E2,G2,D2)+L_{\mathrm{CC2}}(E1,G2,E1,D1) \quad (10.1)$$

where $L_{VAE}$ signifies objective to minimize variational upper bond, $L_{GAN}$ denotes the GAN's objective function, and $L_{CC}$ represents the variational auto encoder's objective function for modeling the cycle-consistency restraint. On MNIST datasets, the UNIT framework outperforms the coupled GAN (Liu & Tuzel, 2016). However, no comparison can be found with another unsupervised approach called CycleGAN discussed by Zhu et al. (2017).

### 10.2.2 EEG Signal Reconstruction and SSVEP Classification

The mutual conflict of good classification performance versus low-cost plagues applications based on electroencephalography inputs. EEG signal rebuilding through high sample rates and sensitivity is difficult due to the nature of this contradiction. Because conventional reconstruction algorithms merely strive to minimize the temporal mean-squared error under general penalties, they lose representative aspects of brain activity and suffer from residual artifacts rather than temporal Mean Squared Error (MSE)based on traditional mathematical approaches. Luo et al. (2020) introduced a novel reconstruction algorithm where the Wasserstein distance (WGAN) and a temporal-spatial-frequency loss function are used in generative adversarial networks. The temporal-spatial frequency MSE-based loss function rebuilds signals using time-series features, common spatial pattern structures, and power spectral density data to compute the MSE. Three motor-related EEG signal datasets having several sensitivities and sampling rates yield promising reconstruction and classification results. Their proposed WGAN architecture increases the average classification accuracy of EEG signal reconstructions with varying sensitivities and the classification of EEG signal reconstructions with the same sensitivity. Figure 10.4 depicts the WGAN design, with the result being an approximate estimate of the Wasserstein-1 distance.

### 10.2.3 Body Sensor-induced Healthcare Applications

Body sensors are becoming increasingly popular for various practical uses, including entertainment, security, and medical research. Wearable sensors can be actively investigated as a means of properly assessing people's health, activities, and behavior. As a result, these

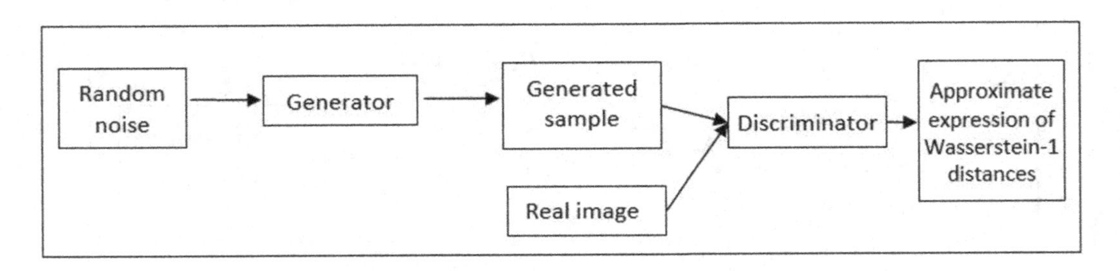


**FIGURE 10.4**
Overview of the WGAN architecture applied in EEG signal reconstruction. Application of Wasserstein distance-based GAN in the reconstruction of electroencephalogram has been illustrated in this figure where the outcome is approximate estimate of the Wasserstein-1 distance.

sensors, like other common electronic gadgets such as computers and cellphones, have the potential to improve our lives. Wearable sensors have primarily been used in commercial applications to trigger panic buttons to seek help in an emergency. This type of sensor application can be deemed a commercial success. The users are assumed to be alert and quick enough to use the button in such instances. In addition, the panic button should be light and pleasant to wear. Wearable sensors have also piqued the interest of many medical scientists who have been studying the physiological functioning of the human body. Patients' vital bodily signals, like heart rate and respiration, are continuously monitored in such apps.

Machine learning models' recent success has been largely attributed to effective deep learning methods having millions of parameters and hundreds of layers. Deep belief network as machine learning technology first succeeded and was later overridden by convolutional neural networks (CNN), specifically for computer vision and image processing applications, among the first deep learning methods. Deep learning based on CNN is particularly good at spotting image patterns, but it hasn't been used much for time-sequential data. Recurrent neural networks (RNN) have been broadly applied to simulate time-sequential events in data in this regard. However, general RNN has a vanishing gradient limitation problem, which happens when long-term information is processed. To address this, the long short-term memory system was created, which contained various memory components. Though the existing deep learning algorithms are effective in their respective sectors, they are susceptible to noise. Hence noise in testing data may reduce the accuracy of overall system performance.

Furthermore, humans are often compelled to accept approaches that are not interpretable, that is, trustworthy, which drives up the need for ethical machine learning by 40–49 percent—focusing solely on model performance rather than describing how decisions gradually lead to system unacceptability. Though there is a trade-off in machine learning between performance and interpretability, increases in explainability can rectify model flaws. As a result, machine learning studies should emphasize creating more understandable models while maintaining high accuracy levels. Figure 10.5 has demonstrated a general outline.

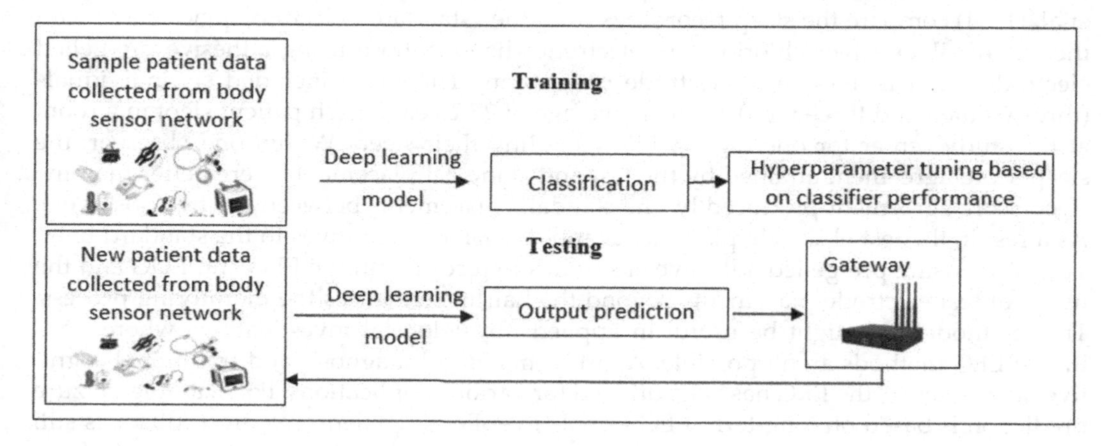


**FIGURE 10.5**
A general outline of body sensor-based healthcare framework using deep learning. A general framework of body sensor-based healthcare system has been presented here where deep learning has played an active role in classification and output prediction.

## 10.3 Healthcare Application Areas of EEG

The previous section discussed the modalities considered in state-of-the-art deep learning-based healthcare systems where EEG was a prominent one used for transformative healthcare. The vital application areas include rare disease diagnosis, robotics-based motor imagery, and rehabilitation from different disorders, which will be discussed in this section.

### 10.3.1 Rare Disease Diagnosis

Based on a recognition concept procedure Van Dun et al. (2009) analyses the real-time achievement of a multiprocessing EEG strategy. Applying eight-channel EEG recordings of 10 standard-hearing subjects, the least electrode collection is identified for auditory steady state responses with reoccurrence ranging from 80 to 110 Hz. In addition, auditory steady state responses with varying frequencies ranging from 80 to 110 Hz in multi-channel processing and originating most often in the brainstem are examined. To acquire the highest average, signal-to-noise ratios (SNRs) and minor change values for independent electrodes, the electrodes must be set behind the head and at the mastoids. In the single-channel processing, auditory steady state responses coming mostly through the auditory cortex with a modulating repetition of 10 Hz are investigated. Individual electrodes placed beyond the head and both mastoids produce the highest mean SNRs. When using a multiprocessing channel, the highest average SNRs are achieved for at least 80 percent of individuals when electrodes are placed at the contralateral mastoid, the forehead F3, and F4, with Cz as a ground source electrode.

Compared to a three-electrode or individual-channel configuration, multi-channel processing is more resistant to errors. In applied physiology, monitoring individuals' sleep in a typical environment (e.g., diving, aerospace) is more commonly required since a reduced quantity or sleep quality might negatively impact work completion. The target of this study was to see whether a simple mechanism of adhesive, pre-gelled electrodes placed to non-hairy regions of the scalp could provide reliable sleep EEG recordings. Here Dyson et al. (1984) compare the sleep recordings from the "standard" electrode placements with the use of silver–silver chloride cup electrodes having those using adhesive pre-gelled electrodes, that is, the "test" electrode placements. The study included six individuals (three women and three men), with an average of 27.2 years. Each patient slept in a room at the study center for one night while recording their sleep. Within one observer, the sleep score agreement attained by the test and standard placements were better than the sleep score agreement produced by the standard placements between the two observers. As a result, the test electrode placements appear viable alternatives to the standard ones. Therefore, using pre-gelled adhesive electrodes to record both the EEG and EOG and the usage of test electrode placements beyond the hairline enhances the monitoring process. This methodology might be useful in applied physiological investigations where traditional EEG methods aren't possible. Apart from clinical diagnosis and traditional cognitive neuroscience, the EEG has been utilized for various applications. For example, seizure prediction is based on long-term EEG signals in epileptic patients. Neurofeedback is still an essential expansion, and it's being used as the foundation for brain–computer interfaces in its most sophisticated form.

In the domain of neuromarketing, the EEG also is popular and widespread. Many commercial items are based on the EEG in some way. For example, Honda is working on a method to control its Asimo robot via EEG, a technology it intends to implement into its cars in the future. Ear-EEG is a discrete and non-invasive method of recording EEG that might be utilized for persistent brain observation in real-world situations. This analysis aimed to establish and analyze dry-contact ear-EEG electrode recordings, which had previously only been done with wet electrodes. A novel ear-EEG platform was created to establish an efficient working of the dry-contact interface. Nano-structured and actively shielded electrodes were placed in a customized soft-earpiece to plan action. In research with 12 participants and four EEG patterns, the platform was examined using alpha-band modulation, mismatch negativity, steady-state auditory response, and steady-state visual evoked potential. The outcome of the dry-connected ear-EEG platform prototype was compared with the recordings of the scalp EEG. So, the dry-contact ear-EEG electrode was shown to be a viable technique for EEG recording in the research undertaken by Kappel et al. (2018).

### 10.3.2 Robotics-based Applications of Deep Learning Inducing EEG

Korovesis et al. (2019) used alpha brain waves as BCI (Kumar et al., 2020); the device is OpenBCI Ganglion, which samples at 200 Hz. They used the infinite impulse response notch filter at 50 Hz, the Butterworth infinite impulse response bandpass filter, and the fast Fourier transform algorithm for signal filtering at 5–15 Hz bandpass. The author uses the multilayer perceptron (MLP) algorithm for classification purposes. Here MLP uses rectified linear unit function as the activation function. They classified the data into four classes, namely, "forward," "reverse," "left," and "right." These classes were used for robotic movements. Their proposed method showed average accuracy of 92.1 percent (Song et al., 2020), proposed another method using P300 as BCI to assist users with disabilities using robots in tasks like grasping something or book turning. They used BrainCap MR with 64 scalp electrodes to capture a 500 Hz sampling rate EEG signals. After the signal acquisition, fourth-order Butterworth bandpass filter linking 1–20 Hz is used. Then the dawn algorithm is used to enhance the signal. The author used linear discriminant analysis, support vector machine, and multilayer perceptron for classification. The author compared their accuracy and showed that support vector machine is preferable at 90.57 percent accuracy overall.

Shedeed et al. (2013) use a four-channel Emotiv Epoc headset device to capture 128 Hz sampling rate EEG signals. They intend to classify three different arm movements: close hand, open arm, and close arm. EEG signals got filtered using a fifth order Butterworth filter. The author uses three techniques: fast Fourier transformation, principal component analysis, and wavelength transformation (WT) for feature extraction. Classification is done using multilayer perceptron using sigmoid function at learning rate of 0.03. They showed different classification rates for different feature extraction techniques. Wavelength transformation shows the best result at 91.1 percent average accuracy for all classes.

Gandhi et al. (2014) acquired an EEG signal at a 256 Hz sampling rate using g.USBamp dry electrode-based system. The author uses recurrent quantum neural network (RQNN) as a feature extraction method. Classification is done using the intelligent adaptive user interface (iAUI), a method proposed by the author. This interface provides two commands every time to the user for movement of the robot to a target point—two-class motor imagery techniques using the left or right hand select either the left or right command. The system's performance is measured using the time and number of commands to complete the task.

Li et al. (2016) used an Emotiv EPOC neuroheadset with a 128 Hz sampling rate to control a robot for signal acquisition. It is a 14-channel wireless device. The author uses an online K-means clustering algorithm for signal processing. After that, principal component analysis is applied for classification to recognize the robotic activity. The robot has four commands: right, left, forward, and backward. At a time, only one command is given to the robot. The author shows an accuracy of 86 percent using this method. Ofner et al. (2019) use movement-related cortical potential for movement prediction, a low-frequency negative shift in EEG signal occurring two seconds before the actual move. Therefore, Magnetic resonance cholangiopancreatography (MRCP) occurs at the time of preparation for some movement. The author uses four 16-channel g.USBamps to collect signals. Sixty-one electrodes got positioned on the scalp; apart from that, three electrodes were used to record electrooculogram (EOG) with a sampling rate at 256 Hz. Signals between 0.3 and 70 Hz are considered. For filtering fourth order, Butterworth filter is used. Then data dimensions are reduced using principal component analysis (PCA). After data processing, shrinkage linear discriminant analysis was applied as the classifier. They used five classes for five hand movements, namely. They achieved 68.4 percent accuracy in classification.

Shao et al. (2020) tried to control a wall-crawling cleaning robot using EEG brain signals. To capture EEG signals, a brain product device is used. Brain product of 500 Hz sampling rate contained 64 electrodes, of which 32 were effective The steady-state visually evoked potential (SSVEP) is used as BCI to communicate with computers. The user selects the robot's target using a flash block on the computer screen. Four different blocks were shown at the screen's right, left, bottom, and top, flashing in different frequencies. After the signal acquisition, wavelength transformation is used as a bandpass filter, and canonical correlation analysis (CCA) is used as a classifier. The mean accuracy was achieved as 89.92 percent using this method. Rashid et al. (2018) used a Neurosky Mindwave EEG headset to capture brain signals at a 512 Hz sampling rate. The author tries to classify if a person is solving quick math or doing nothing. After acquiring a signal, alpha and beta waves from that signal were filtered, representing relaxing and active thinking. For the filtering process, fast Fourier transformation (FFT) is used. The signal is classified next using a machine learning algorithm. Three classification algorithms, namely Linear Discriminant Analysis (LDA), Support Vector Machine (SVM), and k Nearest Neighbor (KNN), were used, among which the best classification accuracy was achieved as 100 percent using SVM.

### 10.3.3 Rehabilitation

It is an activity to assist a person suffering from any disability or illness. He lost his ability partially or completely to do some action or movement. For example, suppose one person cannot speak well after the stroke; using numerous techniques can help him shorten the problem. So, it will be going on till the full recovery of the patient. Motor stultification, such as paralysis, is the most common disorder after the stroke. Motor rehabilitation is a type of practice that is pivoted to recouping a particular function, such as the movement of body parts. It needs specific training by which the patients can learn again and which helps them to rehabilitate.

The design of rehabilitative aid will require seamless interaction between the task executing output device and the BCI operator without cognitive disruption. Based on the review, it has been observed that lower limb tasks involving hip joint and knee movements have not been applied ever in rehabilitative exercise (Deng et al., 2018) since only foot and upper limb kinesthetics has been deployed. EEG-based activity recognition frameworks only consist of orthosis and exoskeletons. No EEG-based prosthetic lower limb or ankle-foot device is currently available as per our knowledge (Roy et al., 2013). Most of the existing applications lack a quantitative performance evaluator to test the applicability of the models in practical scenarios. Factors like walking endurance, energy consumption, and system cost have been ignored in most cases. EEG-based BCI models have mostly applied non-invasive techniques instead of invasive ones and face different constraints due to limited feature size in complex movements. More efficient and robust techniques must be implemented to carry out further research.

The rehabilitation-based healthcare application areas are discussed in this section, where EEG has been extensively applied as the data acquisition device. For absolute and relative power, the number of studies showing a significant increase, reduction, or no significant change in power in each frequency band relative to control for each disorder is indicated in Newson and Thiagarajan (2019). For example, significant declines in absolute power dominated the alpha band for obsessive compulsive disorder (OCD) (eyes closed) and schizophrenia, post-traumatic stress disorder (PTSD), and autism. And in the beta band, it was observed for internet addiction, autism, and attention deficit hyperactivity disorder. On the other hand, significant increases were found in a few diseases, most often. At the same time, individuals' eyes were open, such as depression (beta, eyes open and closed), bipolar (beta and alpha, eyes open), schizophrenia (beta and alpha, eyes open), and alcohol habit (alpha and beta, eyes open) (eyes closed, beta).

#### 10.3.3.1 Bipolar Disorder

It is cerebral sickness that leads to an innumerable low and high frame of mind and substitutes in many activities like sleep, thinking, and feeling. The main character of bipolar disorder is a swing of mood, such as extreme happiness to extreme angriness. Some noticeable disorders are unrestrained excitement, happiness, restlessness, alcohol abuse, less judgment capacity, and reduction of concentration level. EEG mainly quantifies the functions of the human brain. In the case of bipolar disorder, EEG mainly helps to detect and identify the disorder (Sana et al., 2009). EEG assists in the early diagnosis of several disorders in the human brain.

#### 10.3.3.2 Drug Rehabilitation

It is a curing process of a person who is too addicted to alcohol, street drugs, and medicine. Various continuous treatment processes like counseling and meditation help in their recovery. Dependency on the drug is a serious problem in human life. It can result in various ailments of the human body, including the human brain (Volkow et al., 2004). EEG is a vital tool that shows the effects of addiction to the drug in the human brain. At first, the data is collected using EEG, and after the analysis, the rehabilitation process is started. Quantitative EEG alterations in heroin addicts have only been studied in a few research. Low-voltage background movement having diminished alpha rhythm, a rise in beta movement, and a considerable number of low-amplitude theta and delta waves in central regions were observed in more than 70 percent of heroin users during the acute withdrawal stage (Polunima & Davydov, 2004). Different proofs imply that cannabis (tetrahydrocannabinol, marijuana) alters prefrontal brain functionality, resulting in impairments in various sophisticated cognitive functions (Egerton et al., 2006).

#### 10.3.3.3 Gait Rehabilitation

It mainly measures the movement of the human body, observation of eye movement, and working of muscles more accurately for learning the motion of humans. By this, we can detect the disorder of an individual. After that, we can give him medical treatment. Using EEG, we can know muscle activity, which contributes vitally to the cortical control of movements (Tommaso et al., 2015). It also helps to know a person's uniqueness (Van et al., 2004).

#### *10.3.3.4 Vascular Hemiplegia Rehabilitation*

It is a recovery process for a patient who has paralysis of one part of his body. This is mainly caused due to stroke, meningitis, cerebral palsy, and cerebral hemorrhage. There are numerous rehabilitation processes like carrying out exercise, training, relearning, and exertion of muscle. These processes vary from patient to patient (Van Buskirk & Zarling 1951).

#### *10.3.3.5 Dementia*

This is one kind of syndrome where there is degradation in human brain memory, capacity to perform daily works, thinking, and analyzing capacity. It mainly affects aging people. It has mainly three stages: the early, middle, and last stages. EEG helps to detect mainly two types of dementia: AD and VaD, which are mainly cortical. EEG studies the patients who have continuous disorders in the human brain to detect dementia (Durongbhan et al., 2018).

#### *10.3.3.6 Epilepsy*

This is a type of disorder in which nerve cells of the human brain are not working properly. There are mainly two types of epilepsy: one that acts on the entire brain and another that acts on the portion of the brain. These originated because of the withdrawal of alcohol, too much fever, and very low blood sugar. For this disorder, there are no particular treatment processes. It can be handled by medicinal treatment. Epilepsy can be identified using EEG clinically. Many identification processes exist, but EEG is the most nominal (Lasitha & Khan, 2016). It records the signal in different conditions. After that, the analysis is done, and the rehabilitation process is started (Gupta et al., 2021).

## 10.4 Significance of Different Electrode Placement Techniques

The widely recognized electrode placement strategies applied in different areas have been described in this section.

### 10.4.1 10–20 International System

A 10–20 system method was designed to position electrodes on the scalp during EEG recordings. The relative distances between electrodes about the size of the head are indicated by the 10 and 20 in the name. The 10–20 system connects scalp points to cerebral cortex points. The 10–20 system uses a letter followed by a number to signify electrode locations. C, T, O, F, and P are utilized letters. Central, temporal, occipital, frontal, and parietal are the letters that correlate to the different areas of the brain. Except for the center position, these are all brain lobes. The right or left side of the head corresponds to the number in the given location. On the right side of the hemisphere, even numbers are

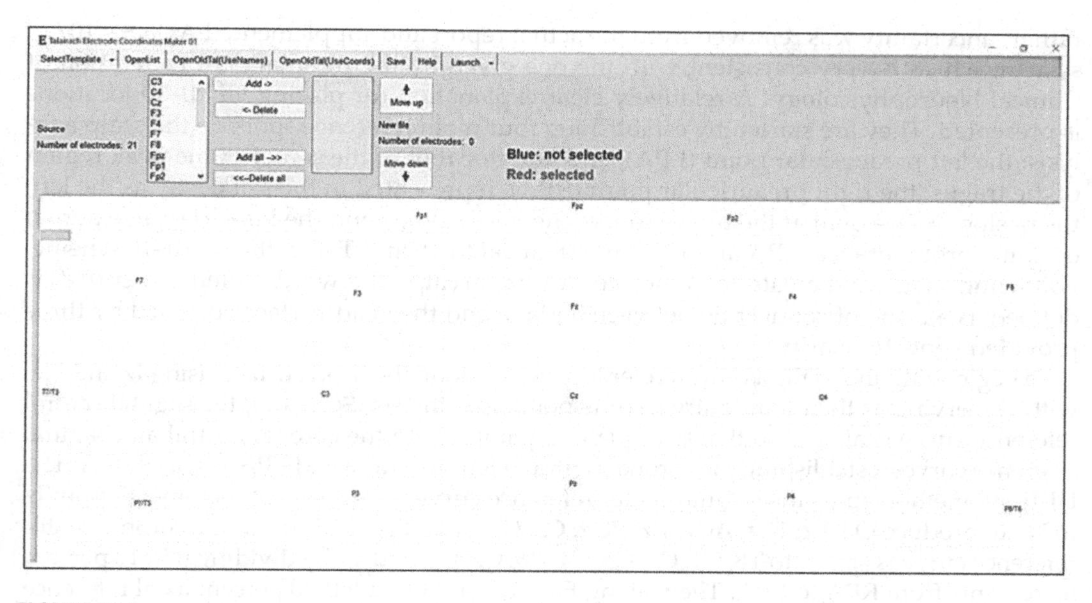


**FIGURE 10.6**
Pictorial view of 10/20 international system in sLoreta. This figure outlines the sLoreta visualization regarding placement of different electrodes on human scalp following the most popular 10/20 International system.

found, whereas odd numbers are found on the left side. Instead of a number, certain letters have a Z after them. The letter Z stands for the head's midpoint. The following criteria for determining the positions are reference points: The inion (the bony mass on the midline at the back of the head near the skull base) and nasion (the upper nose portion in eyes level). The perimeters of the skull in the median planes and transverse are measured from these locations. The positions of the electrodes are selected by splitting these perimeters into 10 percent and 20 percent gaps. Three more electrodes are positioned equidistant from the surrounding spots on each side. RMS EEG monitoring equipment with a 30 mm/s speed captures ECG waveforms. According to the worldwide 10–20 system, a tight-fitting headband was utilized to retain the electrodes in the forehead sites and FP1 and FP3. To guarantee a flat surface and mild pressure, the electrodes were cushioned. Cables with crocodile clips and snap buttons were used to join them. For the remaining places, standard electrodes were utilized. Figure 10.6 shows the electrode locations following the 10–20 international system represented in sLoreta.

### 10.4.2 10–10 System

By introducing a multi-channel EEG device and the subsequent progress of tomographic signal source localization methodologies and topographic techniques, the necessity for expanding the 10–20 system to greater mass electrode setups became more important. As a result, the 10–10 system was suggested, expanding the existing 10–20 system with a greater channel density of 81. There are currently numerous branches and variants of the 10–20 system that are often used without precise explanations. Comparing various derivatives is a bit of an anomaly: there is no clear, typical approach, yet we must deal with derivatives' variety. As a realistic alternative, we can first offer the "unambiguously illustrated (UI) 10/10 system" as a definite standard. This is not an innovation;

rather, uncertainty was removed from the actual report and supplemented ACNS's 10/10 system, which is very consistent with the one given by the International Federation of Clinical Neurophysiology. A relatively clear explanation for placing UI 10–10 locations is presented. They are started by establishing four main reference spots on the scalp analysis: the left preauricular point (LPA), and anterior root of the center of the peak region of the tragus; the right preauricular point (RPA), represented in the same way as the left; the nasion (Nz), a dent at the upper root of the nose bridge; and the inion (Iz), an external occipital protuberance. LPA and RPA are identical to T9 and T10 in the UI 10–10 scheme. Following that, we'll create reference curves on a scalp. The word "reference curve" is defined as a route of connection between a plane and the head surface specified by three provided spots for clarity.

To begin with, the sagittal central reference curve along the head surface using Iz and Nz, with Cz serving as their temporary middle point, is being set. Secondly, the sagittal central reference curve is altered so that Cz equally separates both the coronal central and sagittal reference curves, establishing the coronal central reference curve and LPA, Cz, and RPA. The UI 10–10 method splits the sagittal center reference curve in ten percent increments from Iz to Nz to produce Oz, Fz, FFz, ACz, Pz, POz, Cz, Cpz, and Fpz. Moreover, the coronal center reference curve is split into T8, C1, C3, C5, C4, C6, Cz, C2, and T7 by dividing it in 10 percent increments from RPA to LPA. Then, along Fpz, T7, and Oz, a left 10 percent axial reference curve is created. To establish FT7, F7, AF7, and Fp1, the left anterior quarter of the curve is split by one-fifth increments from Fpz to T7. To obtain O1, PO7, P7, and TP7 for the left posterior quadrant split by one-fifth increments from Oz to T7. For the right hemisphere, the same procedure is used. Following that, six coronal reference curves are established. Starting through four coronal reference curves in the center, using the frontal coronal reference curve as a sample because posterior–occipital (PO) and anterior–frontal (AF) reference curves obey a separate rule. F7, Fz, and F8 are used to describe the frontal coronal reference curve. F5, F3, and F1 are generated by dividing the F7–Fz section of the curve by one-fourth increments from Fz to F7. A similar procedure is repeated on the right hemisphere for the F8–Fz section. The temporal or centroparietal (TP/CP), frontocentral/temporal (FC/FT), and parietal (P) coronal reference curves are divided into quarters on each hemisphere. The anterior–frontal coronal reference curve is calculated using AFz, AF7, and AF8. Because quarterly dividing up causes overcrowding, the AFz–AF7 section of the curve is only divided to produce AF3, and similarly to produce AF4, the AFz–AF8 portion is bisected.

Similarly, PO3 and PO4 are established using the parieto-occipital (PO) coronal reference curve. Finally, along with Iz, LPA (T9), and Nz, we established a left 0 percent axial reference curve. To set F9, FT9, N1, and AF9 for the left anterior quarter, divide by one-fifth increments from Nz to LPA (T9). From LPA (T9) through Iz, divide by one-fifth increments to get TP9, P9, PO9, and I1 for the left posterior quarter. Similarly, set I2, F10, AF10, TP10, FT10, PO10, P10, and N2 for the right hemisphere.

In this manner, 81 locations were selected (eliminating A1 and A2) of the UI 10–10 systems, making every attempt to eliminate any uncertainty. All of the roles specified in ACNS's 10–10 system are included in these 81 positions. In the rest of the study, systems are compared using the UI 10–10 system as the benchmark. Figure 10.7 shows the pictorial view of electrode positions in the 10–10 international system represented in sLoreta.

### 10.4.3 10–5 System

As seen above, the 10–20 and 10–10 methods may create exact and consistent scalp landmarks if the major reference curves and reference points are chosen carefully. The next

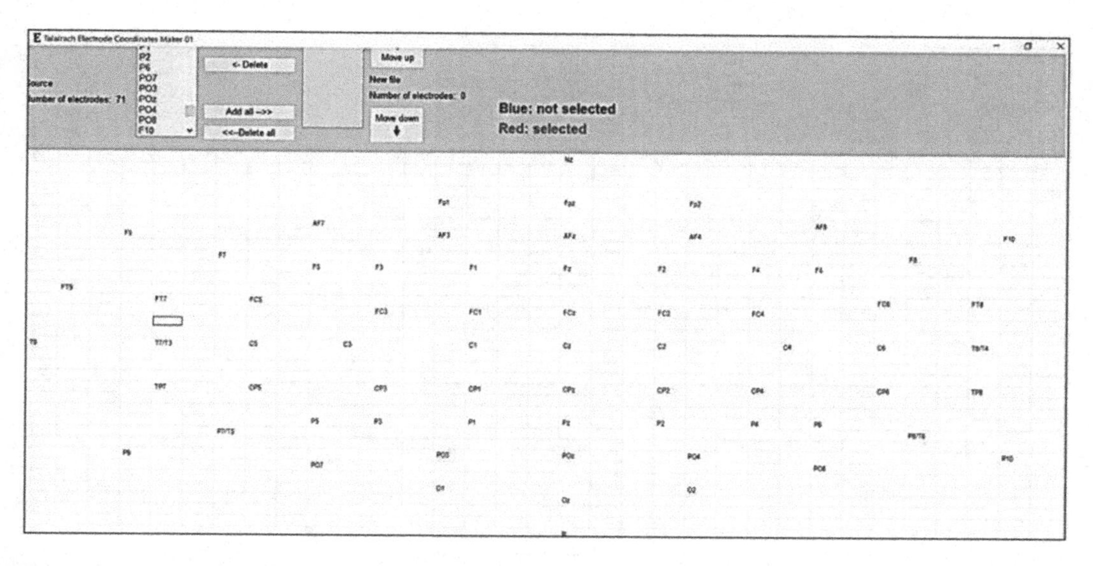


**FIGURE 10.7**
Pictorial view of 10/10 international system in sLoreta. This figure outlines the sLoreta visualization regarding placement of different electrodes on human scalp following the 10/10 International system.

question concerns the system's extensibility. A 10–5 method with more than 300 different scalp landmarks has been put forward so far, however, there is a risk of over-marking. As a result, virtual 10–5 measurements were used on the MR images of 17 participants to investigate the variableness of 10–5 locations for multi-subject research. The measurements were preserved near Oostenveld and Praamstra's original description as feasible (2001). First, the coronal, center sagittal, and 10 percent axial reference curves were drawn as stated above in Oostenveld's 10–10 method. Next, 10–5 standard locations were selected at 5 percent intervals on the central sagittal and 10 percent axial reference curves. The posterior and anterior 5 percent points on 10 percent axial reference curves, which Praamstra and Oostenveld could not fully define, were exceptions to this norm. Extending their reasoning on the 10 percent axial curve, these locations were named Fp1h, Fp2h, O1h, and O2h. NFpz was defined as the point between Nz and Fpz, while NFp1h and NFp2h were labeled as neighboring spots on the five percent axial curve. Next, designers worked on each hemisphere individually on the center coronal curve on the T7–Cz–T8 plane, creating 10–5 typical locations at a 12.5 percent gap between Cz and T7 and continuing the same operation on the right hemisphere. The same procedure was implemented with the central coronal curve for additional coronal reference curves anterior to POO and after AFp. Coronal curves for POO and AFp were drawn, but midway spots were removed. There is no clear explanation for establishing spots on zero percent and 5 percent axial reference curves that we are aware of. Two options independently set the 0 and 5 percent axial reference curves or broadening coronal reference curves dorsally. Nz, LPA, and Iz were used to create a plane and 10–5 standard locations at 5 percent distances on the left hemisphere.

On the other hand, the eyes are likely to have anterior 5, 10, and 15 percent points. Pseudo-T10h and pseudo-T9h 5 percent were placed above the preauricular points from LPA–Cz–RPA for the 5 percent axial curve. A 10/5 and plane standard locations were created at 5 percent distances using Olz, NFpz, and pseudo-T9h. The eye is most likely to have anterior 10, 15, and 20 percent points. The right hemisphere was treated in the

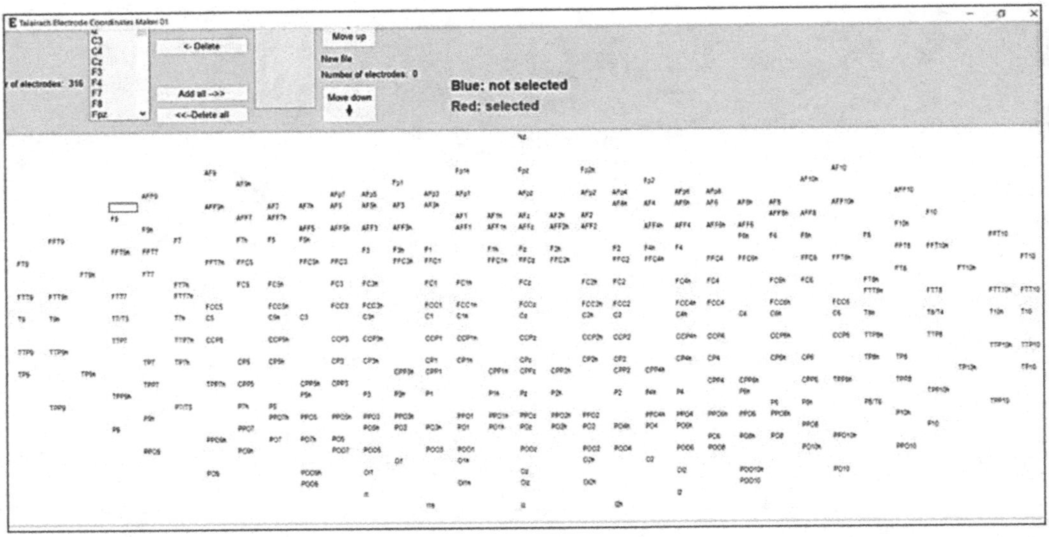

FIGURE 10.8
Pictorial view of 10/5 international system in sLoreta. This figure outlines the sLoreta visualization regarding placement of different electrodes on human scalp following the 10/5 international system.

same way. In this approach, the distribution of 329 Oostenveld's 10/5 standard locations was evaluated (plus 9 extra locations) and elaborated their distribution at the MNI space. Figure 10.8 shows the pictorial view of electrode positions in the 10–5 international system represented in sLoreta.

## 10.5 Conclusion

The current study has discussed the application areas specific to the three widely known electrode placement strategies of EEG. sLORETA visualizations of the techniques have also been included. In addition, the significance and working mechanism of EEG for deep learning-based transformative healthcare applications have been analyzed. Even though conditional generation offers flexibility in better training data resolution and augmentation, just a few research studies exist on the recognition and registration of medical pictures. After reviewing different existing works, we have found that most of the picture translation and reconstruction employ classic metric system methodologies for quantitative evaluation of proposed solutions. When GAN incorporates extra loss, then some difficulty is around in optimizing the visual standard of an image in the absence of a dedicated reference metric, which can be addressed soon. There has been some debate about the rising number of EEG tests or papers claiming superior outcomes using deep learning. When possible, reproducibility and comparisons to existing benchmarks are essential, and their absence should be considered when assessing any claims. Nonetheless, these best practices are becoming more widespread these days, and data publication is compulsory in some circumstances. They are always an excellent sign of the work's quality and a good place to start for projects.

## References

Aznan, N. K. N., Atapour-Abarghouei, A., Bonner, S., Connolly, J. D., Al Moubayed, N., & Breckon, T. P. (2019, July). Simulating brain signals: Creating synthetic EEG data via neural-based generative models for improved ssvep classification. In 2019 International Joint Conference on Neural Networks (IJCNN) (pp. 1–8). IEEE.

de Tommaso, M., Vecchio, E., Ricci, K., Montemurno, A., De Venuto, D., & Annese, V. F. (2015, June). Combined EEG/EMG evaluation during a novel dual task paradigm for gait analysis. In 2015 6th International Workshop on Advances in Sensors and Interfaces (IWASI) (pp. 181–186). IEEE.

Deng, W., Papavasileiou, I., Qiao, Z., Zhang, W., Lam, K.Y. and Han, S., 2018. Advances in automation technologies for lower extremity neurorehabilitation: A review and future challenges. *IEEE Reviews in Biomedical Engineering, 11*, 289–305.

Durongbhan, P., Zhao, Y., Chen, L., Zis, P., De Marco, M., Unwin, Z. C., Venneri, A., He, X., Li, S., Zhao, Y., Blackburn, D. J., & Sarrigiannis, P. G. (2018). A dementia classification framework using frequency and time-frequency features based on EEG signals. IEEE Transactions on Neural Systems and Rehabilitation Engineering (pp. 1534–4320). IEEE.

Dyson, R. J., Thornton, C., & Dore, C. J. (1984). EEG electrode positions outside the hairline to monitor sleep in man. *Sleep, 7*(2), 180–188.

Egerton, A., Allison, C., Brett, R.R., & Pratt, J.A., 2006. Cannabinoids and prefrontal cortical function: insights from preclinical studies. *Neuroscience & Biobehavioral Reviews, 30*(5), 680–695.

Fahimi, F., Zhang, Z., Goh, W. B., Ang, K. K., & Guan, C. (2019, May). Towards EEG generation using GANs for BCI applications. In 2019 IEEE EMBS International Conference on Biomedical & Health Informatics (BHI) (pp. 1–4). IEEE.

Gandhi, V., Prasad, G., Coyle, D., Behera, L., & McGinnity, T. M. (2014). EEG-based mobile robot control through an adaptive brain–robot interface. *IEEE Transactions on Systems, Man, and Cybernetics: Systems, 44*(9), 1278–1285.

Goodfellow, I. J., Pouget-Abadie, J., Mirza, M., Xu, B., Warde-Farley, D., Ozair, S., Courville, A., & Bengio, Y. (2014). Generative adversarial nets. *Advances in Neural Information Processing Systems*, vol. 27.

Gupta A, Chinmay C., & Gupta B. (2021). Secure transmission of EEG data using watermarking algorithm for the detection of epileptical seizures, Traitement du Signal, *IIETA, 38*(2), 473–479.

Hartmann, K. G., Schirrmeister, R. T., & Ball, T. (2018). EEG-GAN: Generative adversarial networks for electroencephalographic (EEG) brain signals. *arXiv preprint arXiv:1806.01875*.

Kappel, S. L., Rank, M. L., Toft, H. O., Andersen, M., & Kidmose, P. (2018). Dry-contact electrode ear-EEG. *IEEE Transactions on Biomedical Engineering, 66*(1), 150–158.

Karthikeyan, M. P., Krishnaveni, K., & Muthumani, N. (2021). Machine Learning technologies in IoT EEG-based healthcare prediction. *Smart Healthcare System Design: Security and Privacy Aspects*, ISBN 1119792258, 1–32.

Kim, T., Cha, M., Kim, H., Lee, J. K., & Kim, J. (2017). Learning to discover cross-domain relations with generative adversarial networks. 34th International Conference on Machine Learning, ICML 2017, *4*, 2941–2949.

Korovesis, N., Kandris, D., Koulouras, G., &Alexandridis, A. (2019). Robot motion control via an EEG-based brain–computer interface by using neural networks and alpha brainwaves. *Electronics, 8*(12), 1387.

Kumar, R.S., Misritha, K., Gupta, B., Peddi, A., Srinivas, K.K., & Chakraborty, C., 2020. A survey on recent trends in brain computer interface classification and applications. *Journal of Critical Reviews, 7*(11), 650–658.

Lasitha S. Vidyaratne, Khan M. Iftekharuddin, "Real-Time Epileptic Seizure Detection Using EEG," 1534–4320 (c) 2016 IEEE.

Li, Y., Zhou, G., Graham, D., &Holtzhauer, A. (2016). Towards an EEG-based brain–computer interface for online robot control. *Multimedia Tools and Applications, 75*(13), 7999–8017.

Liu, M. Y., & Tuzel, O. (2016). Coupled generative adversarial networks. Advances in Neural Information Processing Systems, Nips, pp. 469–477.

Liu, Z., Bicer, T., Kettimuthu, R., Gursoy, D., De Carlo, F., & Foster, I. (2020). TomoGAN: low-dose synchrotron x-ray tomography with generative adversarial networks: discussion. *Journal of the Optical Society of America A*. https://doi.org/10.1364/josaa.375595

Luo, T. J., Fan, Y., Chen, L., Guo, G., & Zhou, C. (2020). EEG signal reconstruction using a generative adversarial network with wasserstein distance and temporal-spatial-frequency loss. *Frontiers in Neuroinformatics*, *14*, 15.

Newson, J.J., & Thiagarajan, T.C., 2019. EEG frequency bands in psychiatric disorders: a review of resting state studies. *Frontiers in Human Neuroscience*, *12*, 521.

Obeid, I., & Picone, J. (2016). The temple university hospital EEG data corpus. *Frontiers in Neuroscience*, *10*, 196.

Ofner, P., Schwarz, A., Pereira, J., Wyss, D., Wildburger, R., & Müller-Putz, G. R. (2019). Attempted arm and hand movements can be decoded from low-frequency EEG from persons with spinal cord injury. *Scientific Reports*, *9*(1), 1–15.

Polunina, A.G. and Davydov, D.M., 2004. EEG spectral power and mean frequencies in early heroin abstinence. *Progress in Neuro-Psychopharmacology and Biological Psychiatry*, *28*(1), 73–82.

Rashid, M., Sulaiman, N., Mustafa, M., Khatun, S., & Bari, B. S. (2018, December). The classification of EEG signal using different machine learning techniques for BCI application. In *International Conference on Robot Intelligence Technology and Applications* (pp. 207–221). Springer, Singapore.

Roy, A.K., Soni, Y., & Dubey, S. (2013) August. Enhancing the effectiveness of motor rehabilitation using Kinect motion-sensing technology. In 2013 IEEE Global Humanitarian Technology Conference: South Asia Satellite (GHTC-SAS) (pp. 298–304). IEEE.

Shao, L., Zhang, L., Belkacem, A. N., Zhang, Y., Chen, X., Li, J., & Liu, H. (2020). EEG-controlled wall-crawling cleaning robot using SSVEP-based brain-computer interface. *Journal of healthcare engineering*, *2020*(6968713), 1–11.

Shedeed, H. A., Issa, M. F., & El-Sayed, S. M. (2013, November). Brain EEG signal processing for controlling a robotic arm. In 2013 8th International Conference on Computer Engineering & Systems (ICCES) (pp. 152–157). IEEE.

Singh, N. K., & Raza, K. (2021). Medical image generation using generative adversarial networks: A review. *Health Informatics: A Computational Perspective in Healthcare*, 77–96.

Song, Y., Cai, S., Yang, L., Li, G., Wu, W., &Xie, L. (2020). A practical EEG-based human-machine interface to online control an upper-limb assist robot. *Frontiers in Neurorobotics*, pp. 14–32.

Van Buskirk, C., & Zarling, V.R., 1951. EEG prognosis in vascular hemiplegia rehabilitation. *AMA Archives of Neurology & Psychiatry*, *65*(6), 732–739.

Van Dun, B., Wouters, J., & Moonen, M. (2009). Optimal electrode selection for multi-channel electroencephalogram based detection of auditory steady-state responses. *The Journal of the Acoustical Society of America*, *126*(1), 254–268.

Van Iersel, M.B., Hoefsloot, W., Munneke, M., Bloem, B.R., & Olde Rikkert, M.G. (2004) Systematic review of quantitative clinical gait analysis in patients with dementia, *Z Gerontol Geriatr*. *37*, 27–32.

Volkow, N.D., Fowler, J.S., & Wang, G.J. (2004). The addicted human brain viewed in the light of imaging studies: brain circuits and treatment strategies. *Neuropharmacology*, *47*, 3–13.

Yasin, S., Asad Hussain, S., Aslan, S., Raza, I., Muzammel, M., & Othmani, A. (2009) EEG based major depressive disorder and bipolar disorder detection using neural networks: A review, arXiv: 13402 [q-bio.NC].

Zhang, A., Su, L., Zhang, Y., Fu, Y., Wu, L., & Liang, S. (2021). EEG data augmentation for emotion recognition with a multiple generator conditional Wasserstein GAN. *Complex & Intelligent Systems*, *8*, 1–13.

Zhu, J. Y., Park, T., Isola, P., & Efros, A. A. (2017). Unpaired image-to-image translation using CycleConsistent adversarial networks. Proceedings of the IEEE International Conference on Computer Vision, 2242–2251. https://doi.org/10.1109/ICCV.2017.244.

# 11

# *Emotion Detection Using Generative Adversarial Network*

**Sima Das and Ahona Ghosh**

**CONTENTS**



## 11.1 Introduction

In today's game development process, content generation is one of the most time-consuming and costly processes. There are two types of game content: functional and non-functional. Non-functional contents, like textures, sprites, and 3D models, are not tied to game procedure and thus have little impact on game dynamics. On the other hand, functional contents like weapons, enemies, and stages are connected to the game procedure and affect game

DOI: 10.1201/9781003203964-11

dynamics in a straightforward way. Levels are crucial in this context, specifically in platform games of first-person shooter type, as they significantly impact the user experience. Unfortunately, the design of the levels actually needs substantial domain knowledge, best practices, and extensive playtesting. Various game academics are exploring and building procedural content generation systems that can model the level creation method and help human designers using machine learning and search techniques for addressing these types of difficulties. In this chapter, current advances of deep learning [1–5] have been reviewed in the context of applying them in playing several kinds of video games like real-time tactic-based games, arcade games, or first-person shooters [6–12]. The specific needs that diverse game genres have on a deep learning framework and open difficulties in applying different machine learning-based approaches to video games, like general gameplay, coping with exceedingly huge decision spaces, and sparse rewards, have been analyzed by us.

The generative structure follows unsupervised learning [13–17], which involves auto-discovery and learning of input patterns. As a result, the model may create or output unique examples that plausibly could have been generated from the actual dataset. There are various applications of GAN in science and technology, including computer vision, security, multimedia and advertisements, image generation, image translation, text to image synthesis, video synthesis, generating high-resolution images, drug discovery, etc. GANs [18–21] are thrilling and fast-moving fields, bringing out the potential of generative models (as shown in Figure 11.1) to create truthful examples found in diverse issue domains, including image-to-image transformation tasks such as converting summer images to winter or day photos to night photos, as well as creating photorealistic object images, locations, and people which even people can't identify as false.

Adaption of the AI [22–24] procedure in gaming is presently a growing research field with various gatherings and committed journals. This chapter audits ongoing developments in deep learning for computer game playing and utilizes game exploration stages though featuring significant to exposed difficulties. The inspiration behind composing this chapter is to survey the area according to the viewpoint of various sorts of games, the challenges they present for deep learning, and the use of deep learning in playing the games. An assortment of audit articles on deep learning exists, just like reviews on support learning and deep support learning, here we center around these strategies adapted to video game playing.

Artificial neural networks (ANNs) [25–28] are broadly useful functions characterized by their structure and the weights of each diagram edge. On account of their over-simplification and capacity to surmise any constant genuine esteem, they take to an assortment of

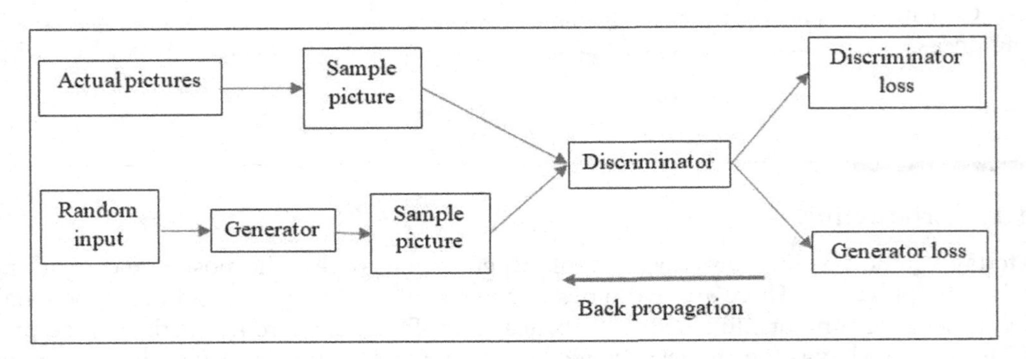


**FIGURE 11.1**
General approach of generative adversarial network.

undertakings, as well as computer game playing. The designs of these ANNs can generally be isolated into two significant classifications: recurrent and feedforward neural networks [29–32]. For instance, the second one takes a solitary contribution, a portrayal of the game state, and yields probabilities or qualities for every conceivable activity. Convolutional neural network [33–36] comprises teachable channels and is appropriate for handling picture information, for example, pixels from a computer game screen. Recurrent neural networks are ordinarily applied to time series information, in which the yield of the organization can rely upon the organization's initiation from past time-step. The preparation cycle is like feedforward networks; then again, the organization's past secret state is taken care of once more and the following information. It permits the organization to become mindful by retaining the past enactments, which is helpful when a solitary perception from a game doesn't address the total game position. For computer game playing, it is normal to utilize a heap of convolutional layers trailed by repetitive layers and completely associated feed-forward layers.

Supervised learning [37–40] in games is used to gain proficiency with the state changes. For example, the neural organization can predict the following state for an activity state pair rather than giving the activity to a given expression. In this manner, the organization learns a game model, improving game playing.

Unsupervised learning targets finding designs in the information rather than learning a plan among information and its names. These calculations can get familiar with the appropriation of components for a dataset, which can group comparative information, pack information into its fundamental elements, or make new manufactured information that is normal for the first information. For games with meager prizes (like Montezuma's Revenge), gaining from information unaided is an expected arrangement and a significant open deep learning test. A noticeable unsupervised learning procedure in deep learning is the auto-encoder. This neural organization endeavors to become familiar with the personality capacity so that the yield is indistinguishable from the info. The principal thought is that the organization needs to figure out how to pack the information and gain proficiency with a decent representation by keeping little. Specialists are starting to use such solo calculations in games to assist with refining high-dimensional information to more important lower-dimensional information; however, this exploration heading is as yet in its beginning phases.

The objective here isn't to foster AI that will make seriously fascinating, dynamic, and sensible game encounters; AI scientists are to a great extent utilizing games as a way to benchmark the insight level of a piece of programming and because virtual universes, with strict guideline and prize frameworks, are an especially valuable climate to prepare to program. The expectation is that human specialists can see how to design machines to perform more confounded undertakings later by encouraging this product to mess around. Deep learning is a type of AI that centers vigorously around the utilization of ANN, which figures out how to tackle complex errands. Deep learning utilizes numerous layers of ANN and different procedures to remove data from info dynamically. Because of this complex, layered methodology, deep learning methods regularly require incredible machines to prepare and run on. The accompanying segments will give a concise outline of distinct enhancement techniques regularly utilized to learn game-playing practices with deep neural organizations. These strategies look for the ideal arrangement of boundaries to take care of some issues. Advancement can likewise be utilized to find hyper-boundaries examined inside deep learning, like organization design and learning boundaries.

The chapter is organized like this. First, the following area outlines various deep learning techniques used in the game designs, trailed by the currently used distinctive exploration stages. Then, Section 11.2 surveys the utilization of DL strategies in various computer game sorts, in Section 11.3, deep learning methods used in gaming applications are surveyed, in Section 11.4, application areas are discussed, and Section 11.5 gives a chronicled outline.

## 11.2 Background Study

This section describes some of the recent works in the related area of research and analyzes them in terms of their benefits and shortcomings to be addressed in future research works. Vries et al. [41] created a power market re-enactment game to help understudies, market experts, and strategy creators comprehend short and long-haul elements. Little gatherings of members in the game assume contending power age organizations. They need to offer their power into a forced trade and choose to put resources into the aging plant. They work in an uncertain climate, not just because they don't have the foggiest idea about their rivals' methodologies, but also because fuel costs, wind, and the accessibility of force plants change. Trendy game, a full business cycle is reproduced, so members experience the impacts of their venture choices. A first augmentation of the game is a $CO_2$ market, which successfully alters the principles of the game halfway. Thus, players figure out how to investigate the effects of strategy tools and strategy vulnerability upon financial backers. Primary involvement of the players demonstrates that the reproduction game makes an undeniable degree of association by members, which results in a lot of further understanding than more conventional pedantic strategies. The further developed agreement that the game gives can likewise be utilized to pass on experiences in related themes, for example, varieties of the arrangement estimate displayed in the game.

Torrado et al. [42] have shown generative adversarial networks (GANs) have provided great outcomes for the picture age. In any case, GANs face difficulties in creating substance with particular sorts of requirements, like game levels. In particular, it is hard to produce levels that have stylish allure and are playable simultaneously. Moreover, because preparation information typically is restricted, it is trying to produce exceptional levels of the existing GANs. In this paper, the authors proposed conditional embedding self-attention GAN (CESAGAN) and a new bootstrapping preparation strategy. It is a change of the self-consideration GAN that consolidates an installing highlight vector contribution to condition the preparation of the generator and discriminator. This mechanism permits the organization to display non-nearby reliance between game items and counting the objects. Moreover, to lessen the number of levels important to prepare the GAN, the authors planned a bootstrapping instrument that adds playable created levels to the preparation set. The outcomes show that the new methodology produces more playable levels and creates fewer copy levels contrasted with a standard GAN.

Anurag et al. [43] investigated mixing levels from existing games to make levels for another game that blends properties of the first games. In this paper, the authors used variational autoencoders (VAEs) for enhancing such strategies. VAEs are fake neural organizations that learn and utilize idle portrayals of datasets to create novel yields. First, they trained a VAE on level information from Super Mario Bros. furthermore, Kid Icarus, empowering it to catch the inactive space spreading over the two games. We then, at that point, utilize this space to produce level portions that join properties of levels from the two games. Additionally, we develop level portions fulfilling explicit imperatives by applying transformative hunt in the dormant space. We contend that these affordances make the VAE-based methodology particularly appropriate for co-imaginative level planning and contrast its exhibition and comparable generative models like the GAN and the VAE-GAN.

Justesen et al. [44] surveyed ongoing deep learning progresses regarding their application in various computer games like first-individual shooters, arcade games, and constant methodology games. Furthermore, they examined the special prerequisites that diverse game classes posture to a deep learning framework. Feature significant open difficulties in applying these AI strategies to video games were also analyzed, like general game playing, managing amazingly huge choice spaces, and meager prizes.

Giacomello et al. [45] developed two generative models to learn from thousand DOOM levels of first-person shooter game and to create new levels from it. Wasserstein GAN having gradient penalty has been employed. Due to the existence of classic DOOM map elements, the majority of produced levels have proven to be interesting to explore and play (like narrow tunnels and large rooms). Because our method pulls domain information through learning, it does not necessitate an expert to explicitly encode it, as traditional procedural generation frequently does. As a result, by putting specific sorts of maps or features in the training set as network inputs, human designers can concentrate on high-level features.

Irfan et al. [46] stated that the deep convolutional generative adversarial networks is an AI approach that can identify how to copy any appropriation of information. Deep convolutional generative adversarial networks are built with a generator and discriminator. The first one creates new substances, and the latter discovers whether the produced content is genuine or counterfeit. Procedural content generation for level age might profit from such models, especially for a game with a current level to imitate. In that paper, deep convolutional GANs are prepared on existing levels produced through an arbitrary generator. Three unique game designs (Colours cape, Zelda, and Freeway) are chosen from the general video game artificial intelligence structure for level age. The proposed framework in the paper effectively produces different levels which imitate the preparation levels. Created levels are additionally assessed utilizing specialist-based testing to guarantee their play capacity.

Oliehoek et al. [47] showed thatGAN had become one of the best systems for solo generative displaying. As GANs are hard to prepare, much exploration has zeroed in on this. Notwithstanding, very little of this examination has straightforwardly taken advantage of game-hypothetical methods. They present generative adversarial network games (GANGs), which expressly model a limited lose–lose situation between a generator ($G$) and classifier ($C$) that utilize blended systems. The size of these games blocks precise arrangement techniques; subsequently, they characterize asset limited best reactions (RBBRs), and an asset limited Nash equilibrium (RB-NE) as a couple of blended systems to such an extent that neither $G$ nor $C$ can track down a superior RBBR. The RB-NE arrangement idea is more extravagant than the thought of "neighborhood Nash equilibria." It catches not just disappointments of getting away from nearby optima of slope plummet, yet applies to any estimated best reaction calculations, incorporating techniques with random restarts. To approve our methodology, they tackle GANGs with the parallel Nash memory calculation, which probably monotonically combines with an RB-NE. They contrast their outcomes with standard GAN arrangements and exhibit that our technique manages common GAN issues, for example, mode breakdown, halfway mode inclusion, and neglect.

In the findings of the research by Park et al. [48], instructive games offer the critical capacity to support customized learning in connecting with virtual universes. In any case, numerous instructive games don't give versatile ongoing interaction to address the issues of individual understudies. To resolve this issue of instructive games, those ought to incorporate game levels that can conform to the particular requirements of individual understudies. Notwithstanding, making countless versatile game levels needs significant exertion by game engineers. A good answer is using procedural substance age to create instructive games that naturally fuse the ideal learning goals. The authors propose a multistep deep convolutional generative antagonistic organization for creating new levels inside a game for software engineering instruction. The model works in two stages: (1) train the generator with a little arrangement of human-wrote model levels and create a much bigger arrangement of engineered levels to increase the preparation information summary generator, and (2) train a subsequent generator utilizing the expanded preparing information and use it to produce novel instructive game levels with improved resolvability. They assess the presentation of the model by looking at the oddity and feasibility of produced levels between the two generators. Output recommends that the proposed multistep model essentially improves the reasonability of the created levels with just minor corruption in the oddity of the produced levels.

Shi et al. [49] showed the game person customization is the center element among numerous new role-playing games where the players can alter the presence of their in-game characters with the corresponding inclinations. The paper concentrates on the issue of consequently making in-game characters with a solitary photograph. In late writing, neural organizations make game motors differentiable, and self-managed learning is utilized to anticipate facial customization boundaries. Be that as it may, in past techniques, the appearance boundaries and facial personality boundaries are deeply combined, making it hard to show the characteristic facial elements of the person. The neural organization-based renderer utilized in past strategies is also hard to reach to multi-see delivering cases. They proposed "PokerFace-GAN" for nonpartisan face game person auto-creation in the above issues. They first formed a differentiable person renderer more adaptable than the past techniques in multi-see delivering scenarios. At that point, they exploit the antagonistic preparation to unravel the appearance boundaries from the personality boundaries viably and subsequently produce player-favored nonpartisan face (demeanor less) characters. All strategy parts can be differentiated so that the technique can be handily prepared under a self-regulated learning worldview. Analysis results prove that the method can create clear, impartial face game characters, which are deeply like the info photographs. The adequacy of the process is confirmed by correlation outcome and removal examinations.

Hald et al. [50] presented a trial way to utilize defined GANs to create levels intended for the riddle game Lily's Garden1. They removed dual state-vectors as of the genuine levels with an end goal to regulate the subtleties of the GAN's yields. Unfortunately, while the GANs perform fine in identifying the top condition (map-shape), they battle to surmise the following condition (piece conveyance). This may be enhanced by evaluating elective designs for the generator and discriminator of the GANs.

Avanaki et al. [51] report on the spilling of interactivity scenes has acquired a lot of consideration, as obvious with the ascent of stages like Twitch.tv, and Facebook Gaming. However, these real-time features need to manage many difficulties because of the bad quality of source materials brought about by customer gadgets, network limits like transmission capacity, and bundle misfortune, just as low postpone prerequisites. Moreover, spatial video antiquity, such as blocking and haziness because of video pressure or up-scaling calculations, can affect end clients' quality of experience of end clients of aloof gaming video web-based applications. In this paper, the authors' research answers to improve the video nature of packed gaming content. As of late, a few super-goal upgrade strategies utilizing GAN like SRGAN have been projected, which are displayed to work with high exactness on non-gaming components. They worked on the SRGAN by adding an altered misfortune work just as changing the generator organization, for example, layer levels and skip associations with work on the progression of data in the organization, which is displayed to develop the apparent quality further. What's more, they present a presentation assessment of further developed SRGAN for the improvement of edge quality brought about by pressure and rescaling ancient rarities for gaming components fixed in numerous goal bitrate sets.

## 11.3 Deep Learning Methods Used in Gaming Applications

This section describes the working mechanisms of different deep learning tools applied in gaming platforms discussed in the existing literature.

### 11.3.1 Super-Resolution GAN

Lower resolution images can be converted to higher resolution using a super-resolution generative adversarial network (SRGAN) as per the requirement in a scenario like a hospital, where storage efficiency is a factor to be considered, and lower resolution images are stored in the patient database. Ancient pieces or historical photos and documents may be invigorated by transferring them into their higher-resolution counterparts and conserved in improved quality. In some of the existing literature [52], the rationale of this design is to recuperate better surfaces from the picture when it gets upscaled so that its quality can't be compromised. There are different strategies, for example, bilinear interpolation, that can be utilized to play out this undertaking; however, they experience the ill effects of picture data misfortune and smoothing. The creators proposed two designs in this paper, one without GAN (SRResNet) and one with GAN (SRGAN). It is presumed that SRGAN has better precision and creates a picture more satisfying to eyes when contrasted with SRGAN. SRGAN has been used to revitalize the old video gaming texture, and in some research, it has been applied.

Like GAN models, the super-resolution GAN likewise contains two sections, generator and discriminator. The generator creates little information depending on the likelihood appropriation, and the discriminator attempts to figure climate information coming from the input dataset or generator. The generator then attempts to advance the created knowledge to trick the discriminator. The generator design contains leftover organizations rather than deep convolution networks. The remaining organizations are not difficult to prepare and permit them to be considerably deeper to produce better outcomes. This is because the leftover organization utilized a kind of association called skip associations. The following are the generator and discriminator compositional subtleties in Figure 11.2. A high-resolution (HR) picture is a down-sample to a low-resolution (LR) picture during the preparation and training. The generator engineering then attempts to up-sample the picture from low goal to super-goal. After that point, the picture is passed into the discriminator. It attempts to recognize a super-goal and high-resolution picture, producing the antagonistic misfortune that gets backpropagated into the generator design.

### 11.3.2 Deep Convolutional Generative Adversarial Network (DC-GAN)

It consists of convolutional-transpose and convolutional layers in the generator and the discriminator, respectively. Finally, Radford et al. proposed unsupervised representation

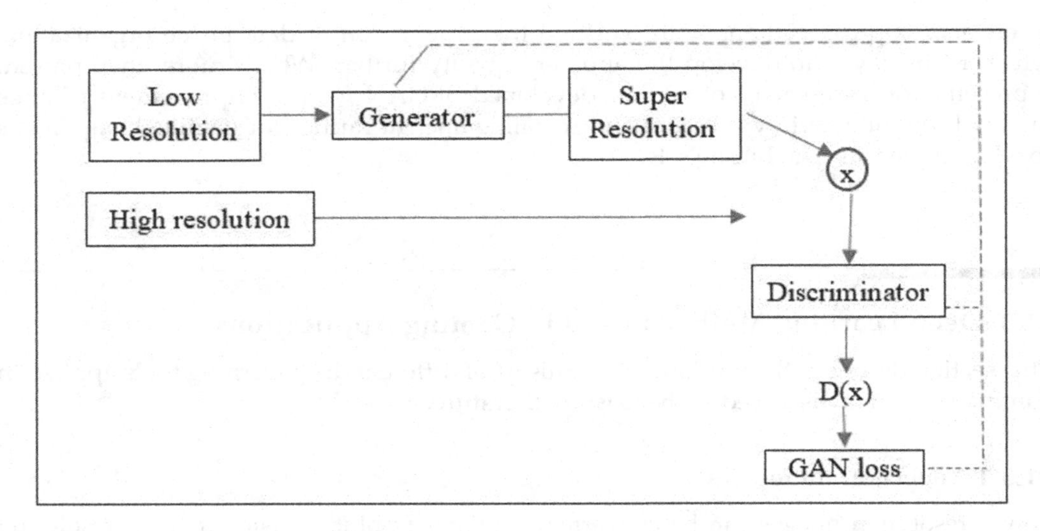


**FIGURE 11.2**
The architecture of super-resolution GAN.

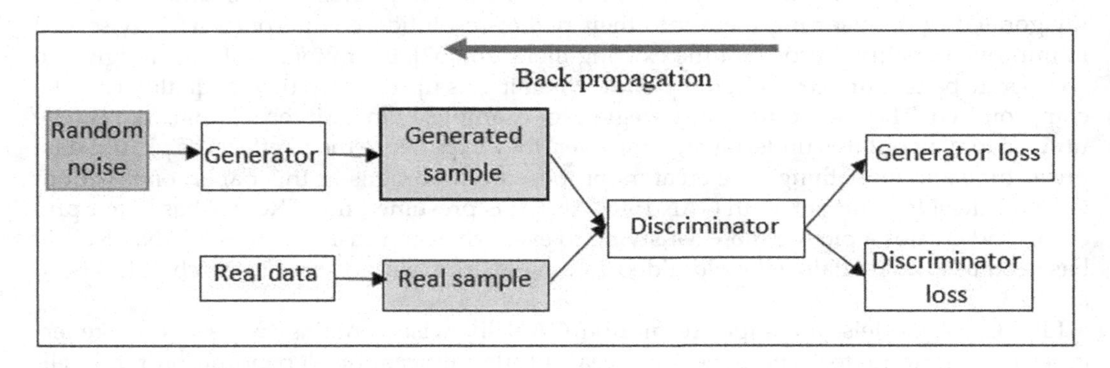


**FIGURE 11.3**
The framework of deep convolutional generative adversarial network.

learning combined with DC-GAN. Here the discriminator consists of stride convolution layers and a group of normalization layers. It takes a 3×64×64 input image. The generator gets created by convolutional-transpose, batch normalization layers, and ReLU initiations. The output will be a 3×64×64 RGB image in Figure 11.3.

### 11.3.3 Conditional Embedding Self-Attention Generative Adversarial Network

It (as shown in Figure 11.4), abbreviated as SAGAN, permits attention-pushed, extended-range dependency modeling for image era obligations. Traditional convolutional GANs generate lower-resolution feature maps of best spatially local factors in lower-resolution feature plots. In the said architecture, particulars may be created applying cues from altogether featuring places. Furthermore, the discriminator may pattern that noticeably distinctive features in remote parts of the image are constant [53–54].

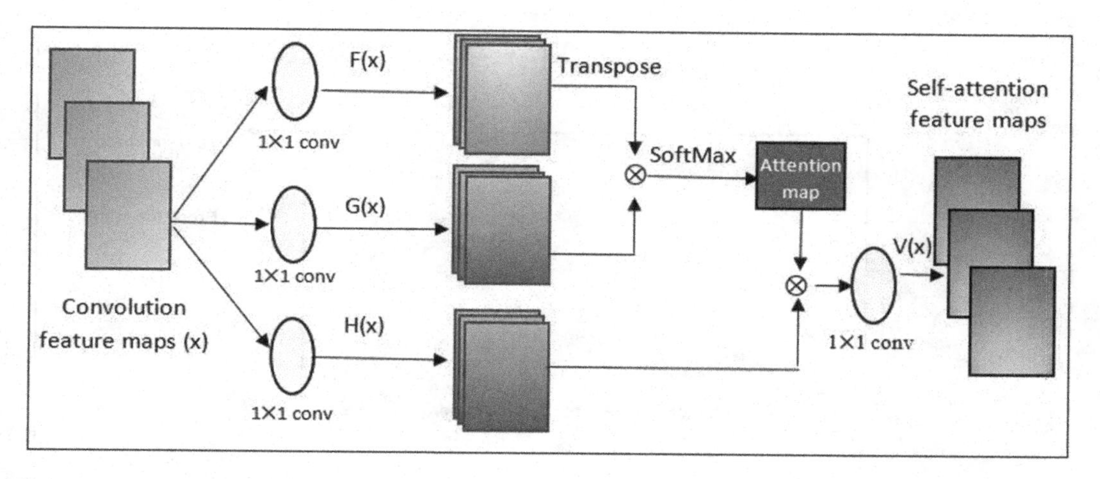


**FIGURE 11.4**
The architecture of conditional embedding self-attention generative adversarial network where ⊗ denotes multiplication of matrix and SoftMax acts as the activation function.

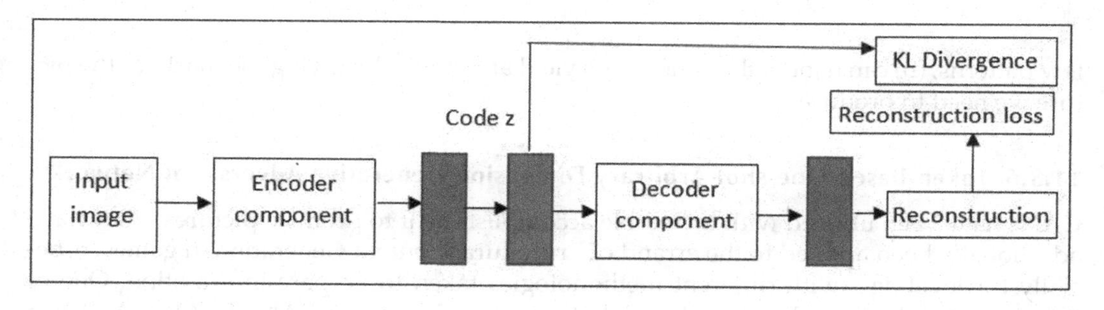


**FIGURE 11.5**
The framework of variational autoencoders generative adversarial network.

### 11.3.4 Variational Autoencoders Generative Adversarial Network

The term variational autoencoders generative adversarial network (VAE-GAN) was first introduced by A. Larsen et al. [55]. The grouping of GANs and variational-autoencoders outperformed the traditional VAEs in the paper (as shown in Figure 11.5).

### 11.3.5 Conditional Generative Adversarial Network (cGAN)

cGAN [56,57] expands the generative adversarial network (GAN), which is utilized as an AI structure for preparing generative models. The thought was first proposed and published by Mehdi Mirza and Simon Osindero in 2014, a paper named "Conditional Generative Adversarial Nets." In cGANs (as shown in Figure 11.6), a restrictive setting is applied, implying that both the generator and discriminator are melded on a type of helper data (class marks or information) from different modalities. Therefore, the ideal model can take in multi-modular planning from contributions to yields by being taken care of with various relevant data. By giving extra information, we get two advantages: (a) convergence will be quicker. Indeed, even the irregular dissemination that the phony pictures follow will have

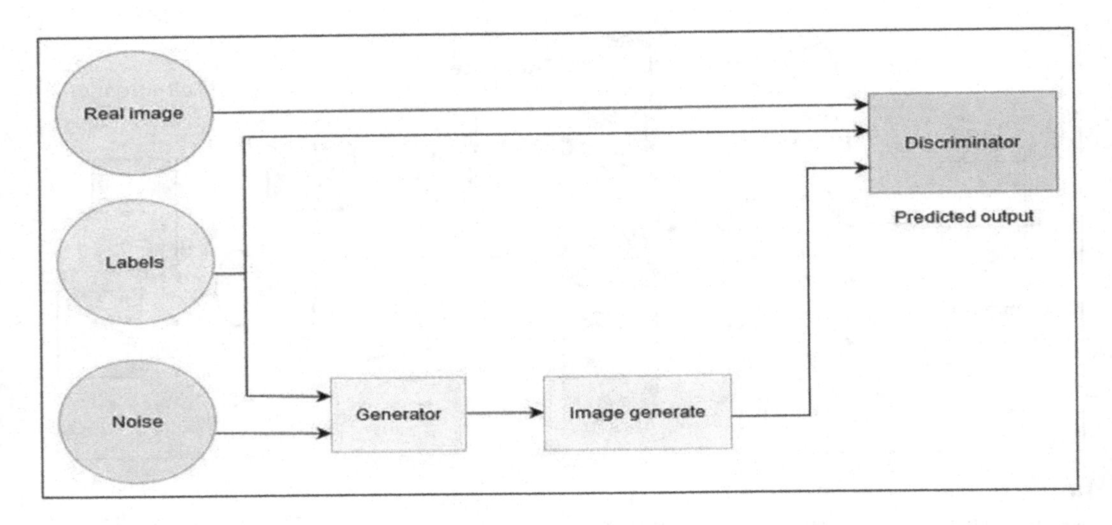


**FIGURE 11.6**
The architecture of conditional generative adversarial network.

few patterns, (b) can handle the generator's yield at test time by giving the mark for the picture we need to produce.

### 11.3.6 Token-based One-shot Arbitrary Dimension Generative Adversarial Network

GANs have been utilized with incredible accomplishment to produce pictures. They have additionally been applied to the errand of Procedural Content Generation in games, especially for level age, with different methodologies taken to prepare information. One of those methodologies, token-based one-shot arbitrary dimension GAN (TOAD-GAN), can create levels dependent on a solitary preparation model and has had the option to duplicate symbolic examples found in the preparation test intently. While TOAD-GAN [58] is an amazing accomplishment, questions stay regarding what precisely it has realized. Can the generator be made to deliver levels that are considerably not the same as the level it has been prepared on? Would it be able to recreate explicit level portions? How unique are the produced levels? The authors examine these inquiries and others by utilizing the CMA-ES calculation for latent space evolution. They utilize an irregular projection in inactive space to make the inquiry space achievable. They also propose the examination embraced here as a worldview for considering what machine-realized generators have learned and a trial of another technique for projecting from a more modest pursuit space to a bigger idle space.

### 11.3.7 Poker-Face Generative Adversarial Network

A devoid demeanor looks described by impartial situating facial provisions, inferring an absence of compelling feeling. It might very well be brought about by an absence of feeling, sorrow, fatigue, or slight disarray, for example, when somebody alludes to something the audience doesn't comprehend. A purposely incited numb demeanor intended to cover one's feelings is otherwise called a poker face, alluding to the normal act of keeping up with one's level-headedness during the playing of game poke. The term emotionless expression was utilized outside the round of poker by American sportswriters during the

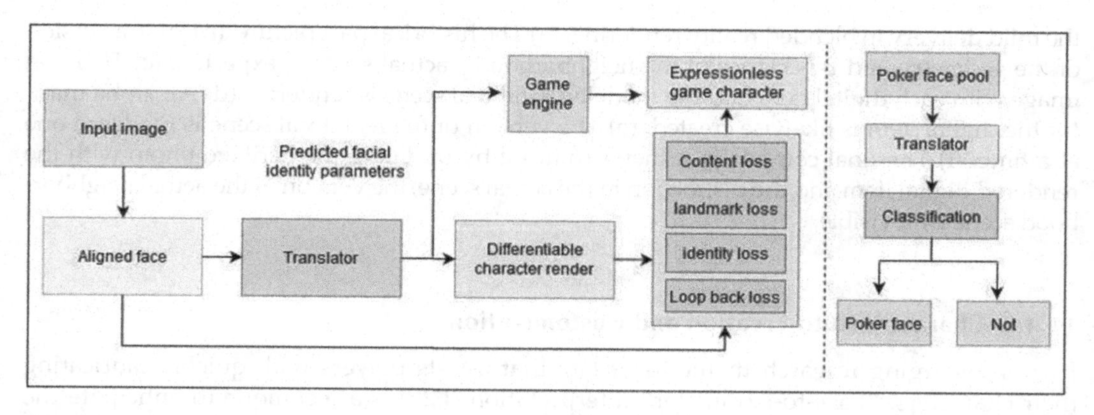


**FIGURE 11.7**
The architecture of poker-face generative adversarial network.

1920s to portray a contender who seemed unaffected by unpleasant circumstances (significant expertise when playing poker for cash, to try not to give a rival any tells around one's hand). It is correspondingly utilized concerning advertisers and sales reps during business arrangements. In Figure 11.7, taking an adjusted facial photograph and foresee three gatherings of facial parameters, a differentiable person renderer that emulates the conduct of the game motor to change the anticipated boundaries over to confront pictures, and a predictor, which is prepared to characterize if the anticipated facial boundaries contain demeanor, are shown. We likewise acquire different misfortune capacities to measure the facial similitude between the delivered and genuine expressions.

## 11.4 Application Areas

Different areas in gaming platforms where GANs have been applied in the state-of-the-art literature have been discussed in this section.

### 11.4.1 Quality Enhancement

Spatial video artifacts such as blockings and haziness as an outcome of up-scaling or video compression procedures can pointedly affect the end-users' quality of experience in passive gaming video streaming applications. In [58], Avanaki et al. investigated results to improve the video resolution of compressed gaming content.

### 11.4.2 Differential Rendering

Deep neural networks (DNNs) have shown fantastic overall performance improvements on vision-related duties, including item detection or photograph segmentation. Despite their achievement, they usually lack the knowledge of 3-D gadgets that form the photograph. It isn't always possible to collect three-D data about the scene or annotate it without problems. It is a unique task that lets the 3D object gradients be designed and propagated through pictures. It also decreases the requirement for 3-D statistics annotation and collection, allowing a higher fulfilment rate in numerous packages. Differential rendering is a popular technique for simulating the mild delivery in blended reality rendering. (a) The historical past picture. (b) First, a version of the geometry and reflectance of the neighborhood's actual scene is expected. (c) Then, an image with each digital item and the modeled local real scene is rendered. (d) An alpha mask for the digital item is likewise created. (e) The version of the real, local scene is rendered one at a time, (f) The final composite is then produced by updating the heritage photo with the rendered digital item and the distinction in the actual scene, the version of the actual neighborhood scene isn't visible.

### 11.4.3 Character Auto-creation and Customization

It is an emerging research theme as of late that assists players with quickly fabricating their characters. "Face-to-parameter" interpretation (F2P) is a technique to anticipate the facial boundaries of in-game characters. In this system, the creators initially train a generative organization to create the game engine differentiable and afterward outline the auto-creation issue as per the neural style move issue. Afterward, a quick and strong strategy adaptation was anticipated depending on self-managed preparation. Additionally, the GANs have been applied to take care of the character auto-creation issue. For instance, Wolf et al. projected a "Tied Output Synthesis" strategy for building defined symbols dependent on an antagonistic preparing structure. In their strategy, the creators initially train a generator to choose a bunch of pre-characterized facial layouts dependent on the information facial photograph and afterward orchestrate the symbol depending on the chosen formats. A typical disadvantage of the above strategies is that these techniques overlook the unraveling of the demeanor boundaries and the facial expression variables. Game character customization is a vital feature of different existing role-playing games, wherein gamers can control the advent of their in-game characters with their options. The paper researches the problem of robotically developing in-recreation characters with a single picture.

In most state-of-the-art research, neural networks are delivered to create game engines differentiable, and the self-supervised gaining knowledge is used to expect facial customization features. Though, in preceding strategies, the expression parameters and facial identity factors are exceedingly coupled, making it hard to model the intrinsic facial capabilities of the person. Apart from this, the neural network-based renderer applied in previous strategies is also tough to be prolonged to multi-view rendering instances. Considering the above problems, this paper recommends a unique technique named "PokerFace-GAN" for impartial face game character auto-introduction. They built a differentiable character renderer more flexibly than the preceding strategies in multi-view rendering cases. Then they take advantage of the antagonistic schooling to disentangle the expression features from the identification features efficiently and, for that reason, generate participant-favored impartial face (expression-much less) characters. As all additives of our technique are differentiable, our technique could be without problems educated below multi-assignment self-supervised gaining knowledge of paradigm. Experiment consequences display that the technique could create shiny neutral face game characters which are tremendously similar to the entered photos.

### 11.4.4 Procedural Content Generation

In computing, it [59,60] is a technique of step-by-step generation of facts in place of manual effort, generally via an aggregate of human-generated belongings to procedures coupled with computer-generated randomness and processing power. Laptop graphics are commonly used to create textures and 3-D fashions. Video games are far applied to automatically make huge quantities of content material in a recreation. Based on the execution, advantages of procedural technology can include smaller report sizes, larger quantities of content material, and unpredictability for fewer probable gameplay. The technological era is a subdivision of mass media combination. Procedural content-technology produces recreation content material, together with stages, searches, or characters, algorithmically.

Motivated with the making of video games playable, reducing authoring problems, and allowing specific aesthetics, several Procedural Content Generation procedures have been developed. Parallelly the investigators are familiarizing strategies from gadget mastering (ML) to procedural content generation troubles; the ML community has become more curious about procedural content generation inspired strategies. One purpose of this development is that ML procedures regularly handiest paintings for various tasks through unique preliminary factors. To tackle the scenario, researchers have started to explore problem parameters randomization to avoid overfitting and allow educated guidelines to be extra effortlessly switched from one surrounding to another, including from a virtual robot to a robot within the practical scenario. They assess present work on procedural content generation, its intersection with modern approaches in ML, and capable new research guidelines and algorithmically generated getting to know surroundings. Although creating in games, they consider procedural content generation algorithms are essential to growing extra popular system intelligence.

### 11.4.5 Video Game Evaluation

In the past years, researchers studying the human–computer interaction network have achieved a huge awareness of growth strategies and strategies used inside the gaming discipline. Affective user-centered design (AUCD) [61] performs a significant position in the game industry since it encourages expressive interaction, subsequently enhancing the interplay modes among customers and video videogames. This paper appears to expand an appropriate AUCD guideline to decide if the stated emotion, semantics, and philosophical idea of a tangible and intangible video gaming interface are nicely obtained using its supposed users. Approaching AUCD in video games calls for examining more than one record to achieve reliable statistics, particularly when evaluating and decoding affect and emotion. They give an assignment because of many ambiguities associated with the useful definition, but it can be unpredictable and complex enough. In this paper, the authors described the techniques and strategies used to evaluate the effective person-centered design in video games. The authors also spoke about the tactics inside the context of current emotional gaming and person-focused layout concept and records collecting techniques, consisting of the elements affecting inside and exterior validity besides the fact's evaluation strategies.

### 11.4.6 User Emotion Identification

It is the system of predicting human emotion. Persons range widely in their recognition of others. The use of era to assist humans with emotional reputation is an exceptionally nascent research region [62,63]. Mainly, technological knowledge works great if it applies numerous sense modalities in context. To date, the maximum work has been carried out on mechanizing the facial expression identification from audio, text, and physiology as per the measurement of the aid of wearables, face expressions, and body gesture-based video and spoken expressions. The graph of emotion detection for analyzing gaming effects is increased [39] with the level of the game increased, as shown in Figure 11.8.

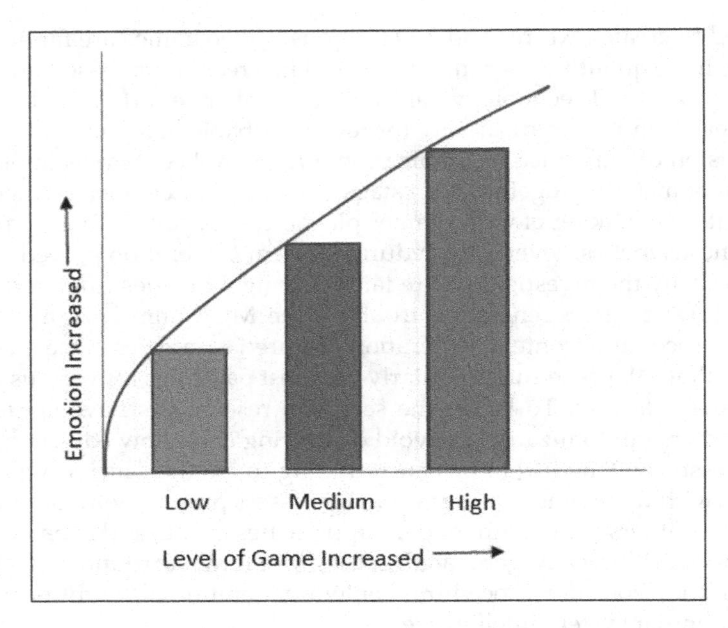


**FIGURE 11.8**
The graph of emotion detection while playing games.

## 11.5 Conclusion

This chapter has described current deep learning methods for playing various computer games. We dissect the unique prerequisites that distinctive game classes posture into some deep learning frameworks and feature significant open difficulties in applying these AI strategies to video games. The greater part of the inspected work is inside, starting to finishing deep model support realizing—convolutional neural organization figures out how to play straight from crude pixels by cooperating through the game. Ongoing work shows that subordinate-free development techniques and hereditary calculations are superior to other options. A portion of the evaluated work involves figuring out how to emulate practices from game logs, while others depend on techniques that gain proficiency with a climate model. In the case of the basic games, the audited strategies can accomplish the above human-computer interaction, although there exist different difficulties in more perplexing games.

In the future, a suitable model with deep learning will be attempted to be developed that can learn with almost no assistance from people, are adaptable to changes in their current circumstances, and can address a wide scope of reflexive and intellectual issues.

## References

1. S. Roy et al., "Deep Learning for Classification and Localization of COVID-19 Markers in Point-of-Care Lung Ultrasound," in IEEE Transactions on Medical Imaging, vol. 39, no. 8, pp. 2676–2687, Aug. 2020, doi: 10.1109/TMI.2020.2994459.

2. H. Tang, H. Liu, W. Xiao, and N. Sebe, "When Dictionary Learning Meets Deep Learning: Deep Dictionary Learning and Coding Network for Image Recognition With Limited Data," in IEEE Transactions on Neural Networks and Learning Systems, vol. 32, no. 5, pp. 2129–2141, May 2021, doi: 10.1109/TNNLS.2020.2997289.
3. S. K. Patnaik, C. N. Babu, and M. Bhave, "Intelligent and Adaptive Web Data Extraction System Using Convolutional and Long Short-term Memory Deep Learning Networks," in Big Data Mining and Analytics, vol. 4, no. 4, pp. 279–297, Dec. 2021, DOI: 10.26599/BDMA.2021.9020012.
4. D. Wu, C. Wang, Y. Wu, Q. -C. Wang and D. -S. Huang, "Attention Deep Model With Multi-Scale Deep Supervision for Person Re-Identification," in IEEE Transactions on Emerging Topics in Computational Intelligence, vol. 5, no. 1, pp. 70–78, Feb. 2021, DOI: 10.1109/TETCI.2020.3034606.
5. Y. Peng, Y. Bao, Y. Chen, C. Wu, C. Meng, and W. Lin, "DL2: A Deep Learning-Driven Scheduler for Deep Learning Clusters," in IEEE Transactions on Parallel and Distributed Systems, vol. 32, no. 8, pp. 1947–1960, 1 Aug. 2021, doi: 10.1109/TPDS.2021.3052895.
6. D. Herumurti, A. A. Yunanto, G. A. Senna, I. Kuswardayan and S. Arifiani, "Development of First-Person Shooter Game with Survival Maze Based on Virtual Reality," 2020 6th Information Technology International Seminar (ITIS), 2020, pp. 81–86, DOI: 10.1109/ITIS50118.2020.9321076.
7. P. B. S. Serafim, Y. L. B. Nogueira, C. Vidal, and J. Cavalcante-Neto, "On the Development of an Autonomous Agent for a 3D First-Person Shooter Game Using Deep Reinforcement Learning," in 2017 16th Brazilian Symposium on Computer Games and Digital Entertainment (SBGames), 2017, pp. 155–163, DOI: 10.1109/SBGames.2017.00025.
8. S. Vlahovic, M. Suznjevic, and L. Skorin-Kapov, "Challenges in Assessing Network Latency Impact on QoE and In-Game Performance in VR First Person Shooter Games," in 2019 15th International Conference on Telecommunications (ConTEL), 2019, pp. 1–8, DOI: 10.1109/ConTEL.2019.8848531.
9. A. Friedman and J. Schrum, "Desirable Behaviors for Companion Bots in First-Person Shooters," in 2019 IEEE Conference on Games (CoG), 2019, pp. 1–8, DOI: 10.1109/CIG.2019.8848036.
10. T. Alves, S. Gama, and F. S. Melo, "Towards Guidelines for Mental State Induction in First-Person Shooters," in 2018 International Conference on Graphics and Interaction (ICGI), 2018, pp. 1–4, DOI: 10.1109/ITCGI.2018.8602700.
11. D. Piergigli, L. A. Ripamonti, D. Maggiorini, and D. Gadia, "Deep Reinforcement Learning to Train Agents in a Multiplayer First Person Shooter: some Preliminary Results," in 2019 IEEE Conference on Games (CoG), 2019, pp. 1–8, DOI: 10.1109/CIG.2019.8848061.
12. A. L. Cricenti and P. A. Branch, "A Generalized Prediction Model of First-person Shooter Game Traffic," in 2009 IEEE 34th Conference on Local Computer Networks, 2009, pp. 213–216, DOI: 10.1109/LCN.2009.5355165.
13. A. Ahmed, I. Zualkernan and H. Elghazaly, "Unsupervised Clustering of Skills for an Online Learning Platform," in 2021 International Conference on Advanced Learning Technologies (ICALT), 2021, pp. 200–202, DOI: 10.1109/ICALT52272.2021.00066.
14. M. A. Kabir and X. Luo, "Unsupervised Learning for Network Flow Based Anomaly Detection in the Era of Deep Learning," in 2020 IEEE Sixth International Conference on Big Data Computing Service and Applications (BigDataService), 2020, pp. 165–168, DOI: 10.1109/BigDataService49289.2020.00032.
15. R. Nian, G. Ji, and M. Verleysen, "An Unsupervised Gaussian Mixture Classification Mechanism Based on Statistical Learning Analysis," in 2008 Fifth International Conference on Fuzzy Systems and Knowledge Discovery, 2008, pp. 14–18, DOI: 10.1109/FSKD.2008.333.
16. P. Du, X. Lin, X. Pi, and X. Wang, "An Unsupervised Learning Algorithm for Deep Recurrent Spiking Neural Networks," in 2020 11th IEEE Annual Ubiquitous Computing, Electronics & Mobile Communication Conference (UEMCON), 2020, pp. 0603–0607, DOI: 10.1109/UEMCON51285.2020.9298074.

17. Xinxin Bai, Gang Chen, Zhonglin Lin, Wenjun Yin, and Jin Dong, "Improving Image Clustering: An Unsupervised Feature Weight Learning Framework," in 2008 15th IEEE International Conference on Image Processing, 2008, pp. 977–980, DOI: 10.1109/ICIP.2008.4711920.
18. X. Yan, B. Cui, Y. Xu, P. Shi, and Z. Wang, "A Method of Information Protection for Collaborative Deep Learning under GAN Model Attack," in IEEE/ACM Transactions on Computational Biology and Bioinformatics, vol. 18, no. 3, pp. 871–881, 1 May-June 2021. DOI: 10.1109/TCBB.2019.2940583.
19. Y. Hua, R. Li, Z. Zhao, X. Chen, and H. Zhang, "GAN-Powered Deep Distributional Reinforcement Learning for Resource Management in Network Slicing," in IEEE Journal on Selected Areas in Communications, vol. 38, no. 2, pp. 334–349, Feb. 2020, doi: 10.1109/JSAC.2019.2959185.
20. L. Chen, R. Wang, D. Yan, and J. Wang, "Learning to Generate Steganographic Cover for Audio Steganography Using GAN," in IEEE Access, vol. 9, pp. 88098–88107, 2021, doi: 10.1109/ACCESS.2021.3090445.
21. S. Wen, W. Tian, H. Zhang, S. Fan, N. Zhou, and X. Li, "Semantic Segmentation Using a GAN and a Weakly Supervised Method Based on Deep Transfer Learning," in IEEE Access, vol. 8, pp. 176480–176494, 2020, DOI: 10.1109/ACCESS.2020.3026684.
22. C. Tang, Z. Wang, X. Sima, and L. Zhang, "Research on Artificial Intelligence Algorithm and Its Application in Games," in 2020 2nd International Conference on Artificial Intelligence and Advanced Manufacture (AIAM), 2020, pp. 386–389, DOI: 10.1109/AIAM50918.2020.00085.
23. A. S. Ahmad, "Brain-inspired Cognitive Artificial Intelligence for Knowledge Extraction and Intelligent Instrumentation System," in 2017 International Symposium on Electronics and Smart Devices (ISESD), 2017, pp. 352–356, DOI: 10.1109/ISESD.2017.8253363.
24. N. Wang, Y. Liu, Z. Liu, and X. Huang, "Application of Artificial Intelligence and Big Data in Modern Financial Management," in 2020 International Conference on Artificial Intelligence and Education (ICAIE), 2020, pp. 85–87, DOI: 10.1109/ICAIE50891.2020.00027.
25. Leon Reznik, "Computer Security with Artificial Intelligence, Machine Learning, and Data Science Combination," in Intelligent Security Systems: How Artificial Intelligence, Machine Learning and Data Science Work For and Against Computer Security, IEEE, 2022, pp. 1–56, DOI: 10.1002/9781119771579.ch1.
26. B. Schrenk, "Simplified Coherent Synaptic Receptor for Filterless Optical Neural Networks," in IEEE Journal of Selected Topics in Quantum Electronics, vol. 28, no. 2, pp. 1–7, March–April 2022, Art no. 7400107, doi: 10.1109/JSTQE.2021.3108573.
27. X. Xu, T. Gao, Y. Wang, and X. Xuan, "Event Temporal Relation Extraction with Attention Mechanism and Graph Neural Network," in Tsinghua Science and Technology, vol. 27, no. 1, pp. 79–90, Feb. 2022, doi: 10.26599/TST.2020.9010063.
28. D. M. Le, M. L. Greene, W. A. Makumi and W. E. Dixon, "Real-Time Modular Deep Neural Network-Based Adaptive Control of Nonlinear Systems," in IEEE Control Systems Letters, vol. 6, pp. 476–481, 2022, doi: 10.1109/LCSYS.2021.3081361.
29. W. Liu, N. Mehdipour and C. Belta, "Recurrent Neural Network Controllers for Signal Temporal Logic Specifications Subject to Safety Constraints," in IEEE Control Systems Letters, vol. 6, pp. 91–96, 2022, doi: 10.1109/LCSYS.2021.3049917.
30. A. N. Sadon, S. Ismail, N. S. Jafri, and S. M. Shaharudin, "Long Short-Term vs Gated Recurrent Unit Recurrent Neural Network For Google Stock Price Prediction," in 2021 2nd International Conference on Artificial Intelligence and Data Sciences (AiDAS), 2021, pp. 1–5, DOI: 10.1109/AiDAS53897.2021.9574312.
31. X. Yang and C. Bao, "A New Four-Channel Speech Coding Method Based on Recurrent Neural Network," in 2021 IEEE International Conference on Signal Processing, Communications and Computing (ICSPCC), 2021, pp. 1–5, DOI: 10.1109/ICSPCC52875.2021.9564779.

32. X. Mo, Y. Xing, and C. Lv, "Graph and Recurrent Neural Network-based Vehicle Trajectory Prediction for Highway Driving," in 2021 IEEE International Intelligent Transportation Systems Conference (ITSC), 2021, pp. 1934–1939, DOI: 10.1109/ITSC48978.2021.9564929.
33. F. A. Uçkun, H. Özer, E. Nurbaş, and E. Onat, "Direction Finding Using Convolutional Neural Networks and Convolutional Recurrent Neural Networks," in 2020 28th Signal Processing and Communications Applications Conference (SIU), 2020, pp. 1–4, DOI: 10.1109/SIU49456.2020.9302448.
34. R. Xin, J. Zhang, and Y. Shao, "Complex Network Classification with Convolutional Neural Network," in Tsinghua Science and Technology, vol. 25, no. 4, pp. 447–457, Aug. 2020, DOI: 10.26599/TST.2019.9010055.
35. B. J. Moore, T. Berger, and D. Song, "Validation of a Convolutional Neural Network Model for Spike Transformation Using a Generalized Linear Model," in 2020 42nd Annual International Conference of the IEEE Engineering in Medicine and Biology Society (EMBC), 2020, pp. 3236–3239, DOI: 10.1109/EMBC44109.2020.9176458.
36. G. Lou and H. Shi, "Face Image Recognition Based on Convolutional Neural Network," in China Communications, vol. 17, no. 2, pp. 117–124, Feb. 2020, DOI: 10.23919/JCC.2020.02.010.
37. O. Baykal and F. N. Alpaslan, "Supervised Learning in Football Game Environments Using Artificial Neural Networks," in 2018 3rd International Conference on Computer Science and Engineering (UBMK), 2018, pp. 110–115, DOI: 10.1109/UBMK.2018.8566428.
38. C. O. Sosa Jimenez, H. G. A. Mesa, G. Rebolledo-Mendez, and S. de Freitas, "Classification of Cognitive States of Attention and Relaxation Using Supervised Learning Algorithms," in 2011 IEEE International Games Innovation Conference (IGIC), 2011, pp. 31–34, DOI: 10.1109/IGIC.2011.6115125.
39. Sima Das, Lidia Ghosh and Sriparna Saha. ( "Analyzing Gaming Effects on Cognitive Load Using Artificial Intelligent Tools," 2020. 10.1109/CONECCT50063.2020.9198662.
40. Sima Das and Aakashjit Bhattacharya. . "ECG Assess Heartbeat Rate, Classifying Using BPNN While Watching Movie and Send Movie Rating through Telegram," 2021. 10.1007/978-981-15-9774-9_43.
41. L. J. De Vries, E. Subramahnian and E. J. L. Chappin, "A Power Game: simulating the Long-term Development of an Electricity Market in a Competitive Game," in 2009 Second International Conference on Infrastructure Systems and Services: Developing 21st Century Infrastructure Networks (INFRA), 2009, pp. 1–6, DOI: 10.1109/INFRA.2009.5397876.
42. Ruben Rodriguez Torrado, Ahmed Khalifa, Michael Cerny Green, Niels Justesen, Sebastian Risi and Julian Togelius. "Bootstrapping Conditional GANs for Video Game Level Generation," in 2020 IEEE Conference on Games (CoG), 2020, pp. 41–48.
43. Anurag Sarkar, Zhihan Yang and Seth Cooper. "Controllable Level Blending between Games using Variational Autoencoders," ArXiv abs/2002.11869, 2020, n. pag.
44. N. Justesen, P. Bontrager, J. Togelius and S. Risi, "Deep Learning for Video Game Playing," in IEEE Transactions on Games, vol. 12, no. 1, pp. 1–20, March 2020, DOI: 10.1109/TG.2019.2896986.
45. E. Giacomello, P. L. Lanzi and D. Loiacono, "DOOM Level Generation Using Generative Adversarial Networks," in 2018 IEEE Games, Entertainment, Media Conference (GEM), 2018, pp. 316–323, DOI: 10.1109/GEM.2018.8516539.
46. A. Irfan, A. Zafar, and S. Hassan, "Evolving Levels for General Games Using Deep Convolutional Generative Adversarial Networks," in 2019 11th Computer Science and Electronic Engineering (CEEC), 2019, pp. 96–101, DOI: 10.1109/CEEC47804.2019.8974332.
47. Frans Oliehoek, Rahul Savani, Jose Gallego-Posada, Elise van der Pol, Edwin de Jong, and Roderich Groß,. GANGs: Generative Adversarial Network Games. arXiv preprint arXiv:1712.00679.
48. K. Park, B. W. Mott, W. Min, K. E. Boyer, E. N. Wiebe, and J. C. Lester, "Generating Educational Game Levels with Multistep Deep Convolutional Generative Adversarial Networks," in 2019 IEEE Conference on Games (CoG), 2019, pp. 1–8, DOI: 10.1109/CIG.2019.8848085.

49. Tianyang Shi , Zhengxia Zou , Xinhui Song , Zheng Song, Changjian Gu, Changjie Fan, and Yi Yuan. "Neutral Face Game Character Auto-Creation via PokerFace-GAN," 2020, pp. 3201–3209. 10.1145/3394171.3413806.
50. Andreas Hald, Jens Hansen, Jeppe Kristensen, and Paolo Burelli.. "Procedural Content Generation of Puzzle Games using Conditional Generative Adversarial Networks," 2020, pp. 1–9. 10.1145/3402942.3409601.
51. N. J. Avanaki, S. Zadtootaghaj, N. Barman, S. Schmidt, M. G. Martini, and S. Möller, "Quality Enhancement of Gaming Content using Generative Adversarial Networks," in 2020 Twelfth International Conference on Quality of Multimedia Experience (QoMEX), 2020, pp. 1–6, DOI: 10.1109/QoMEX48832.2020.9123074.
52. M. Rajabi, M. Ashtiani B. MinaeiBidgoli and O. Davoodi "A Dynamic Balanced Level Generator for Video Games Based on Deep Convolutional Generative Adversarial Networks," Scientia Iranica, vol. 28 (Special issue on collective behavior of nonlinear dynamical networks), pp.1497–1514, 2021.
53. N.J. Avanaki,S. Zadtootaghaj,N. Barman,S. Schmidt,M.G. Martini and S. Möller "Quality Enhancement of Gaming Content Using Generative Adversarial Networks," in 2020 Twelfth International Conference on Quality of Multimedia Experience (QoMEX), 2020, pp. 1–6. IEEE.
54. N. Justesen,P. Bontrager,J. Togelius S. and Risi "Deep Learning for Video Game Playing," IEEE Transactions on Games, vol. 12(1), pp.1–20, 2019.
55. A.B.L. Larsen, , S.K. Sønderby, , H. Larochelle, and O. Winther. June. "Autoencoding Beyond Pixels Using a Learned Similarity Metric," in International Conference on Machine Learning, 2016, pp. 1558–1566. PMLR.
56. L. Fan, Q. Yang, B. Deng, Y. Zeng, and H. Wang, "Concealed Object Detection For Active Millimeter-Wave Imaging Based CGAN Data Augmentation," in 2021 14th UK-Europe-China Workshop on Millimetre-Waves and Terahertz Technologies (UCMMT), 2021, pp. 1–3, DOI: 10.1109/UCMMT53364.2021.9569893.
57. L. Yang, "Conditional Generative Adversarial Networks (CGAN) for Abnormal Vibration of Aero Engine Analysis," in 2020 IEEE 2nd International Conference on Civil Aviation Safety and Information Technology (ICCASIT), 2020, pp. 724–728, DOI: 10.1109/ICCASIT50869.2020.9368622.
58. F. Schubert, M. Awiszus and B. Rosenhahn, "TOAD-GAN: a Flexible Framework for Few-Shot Level Generation in Token-Based Games," in IEEE Transactions on Games, DOI: 10.1109/TG.2021.3069833.
59. R.R. Torrado, , A. Khalifa, , M.C. Green, , N. Justesen, , S. Risi, and J. Togelius, "Bootstrapping Conditional Gans for Video Game Level Generation," in 2020 IEEE Conference on Games (CoG), 2020, pp. 41–48. IEEE.
60. J. P. A. Campos and R. Rieder, "Procedural Content Generation using Artificial Intelligence for Unique Virtual Reality Game Experiences," in 2019 21st Symposium on Virtual and Augmented Reality (SVR), 2019, pp. 147–151, DOI: 10.1109/SVR.2019.00037.
61. Y. Y. Ng and C. W. Khong, "A Review of Affective User-centered Design for Video Games," in 2014 3rd International Conference on User Science and Engineering (i-USEr), 2014, pp. 79–84, DOI: 10.1109/IUSER.2014.7002681.
62. A. Ghosh and S. Dey, "'Sensing the Mind': An Exploratory Study About Sensors Used in E-Health and M-Health Applications for Diagnosis of Mental Health Condition," in *Efficient Data Handling for Massive Internet of Medical Things*, 2021, pp. 269–292. Springer, Cham.
63. K. Chanda, A. Ghosh, S. Dey, R. Bose, and S. Roy, "Smart Self-Immolation Prediction Techniques: An Analytical Study for Predicting Suicidal Tendencies Using Machine Learning Algorithms," in *Smart IoT for Research and Industry*, 2022, pp. 69–91. Springer, Cham.

# 12

# *Underwater Image Enhancement Using Generative Adversarial Network*

**Nisha Singh Gaur, Mukesh D. Patil, and Gajanan K. Birajdar**

**CONTENTS**



## 12.1 Introduction

It's often observed many the times, sea exploration has now gained importance as humans have always been always curious about how we know very little about the underwater environment, even if the sea covers almost 71 percent of the earth. Even when humans try to explore the resources underwater, it has been evident that observations made underwater are not always the same, they always vary based on various parameters such as molecules present in water at the time of data collection, depth of the water when remotely operated vehicle (ROV) or autonomous underwater vehicle (AUV) are collecting data [1–3].

The impact of light propagation in underwater is non-linear. The underwater visual information is often collected at such a depth that where the depth is often dominated by a green or blue hue, whereas the red wavelength gets absorbed if we go deeper into the water as shown in Figure 12.1. This kind of wavelength-posed attenuation may cause the distortion, which in terms is non-linear and the same can be observed through the data gathered underwater.

DOI: 10.1201/9781003203964-12

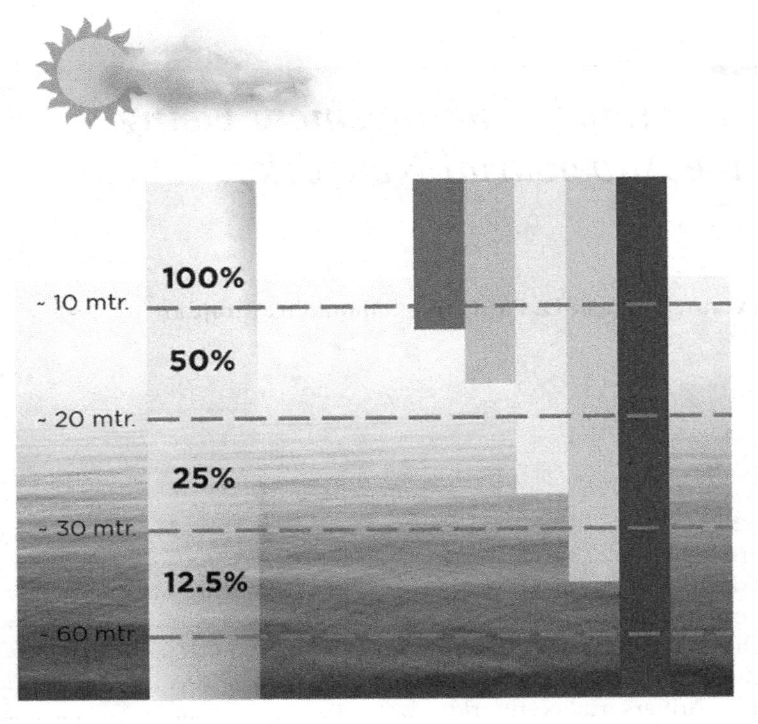


FIGURE 12.1
Light at various depths in the underwater environment.

Even various works of literature have followed various methods to model these issues. Some of the pieces of literature have followed the learning-based approach, which in fact has presented good results. Due to this improvisation quality, underwater visual data can be obtained. In model-based approaches, convolutional neural network (CNN) and GAN-based approaches provide remarkable outcomes in underwater imagery enhancement in terms of colour enhancement, contrast improvement and de-hazing of images have been addressed and solutions for the same have been proposed [1–7]. Yet there is scope for improvising and designing an efficient approach or the model for enhancement of underwater image/video. Also, there is room for formulating the solution for the real-time underwater image/video enhancement, which has not been considered in the literature review of this chapter.

## 12.2 Literature Review

In the recent decade, tremendous research is going on in the area of underwater image/video enhancement, and out of various techniques proposed, most of them nowadays are based on deep learning, CNN, and GAN.

Wang et. al [1] presented the network based on CNN called UIE-Net, which performs the function of colour correction and haze removal, based on synthesized images, generated on an underwater imaging model and was able to achieve good accuracy over the existing methods.

Li et al. [2] demonstrated the results which showed the improvisation in terms of mean average precision (MAP), which is relatively better than that of the other methods for fish detection and recognition methods. CNN is widely used in an underwater environment for object detection, image classification and motion classification as it provides better accuracy over traditional methods employed for the same task.

In 2020, Han et al. [3] used the deep CNN for detection and classification in an underwater environment, various methods were used to compare the results like YOLO, Fast RCNN, Faster RCNN, etc. it is evident that the results achieved for higher speed of detection are better and also the mAP is about 90 percent, which could help in real-time detection using marine robotics. Due to the lack of dataset in the underwater environment, sometimes it becomes necessary to have a synthetic dataset that can be generated through the use of GAN; it learns the relationship between underwater images which are degraded and the images with fewer imperfections or clear images. With this method, the images were restored by training the model.

In 2014, Goodfellow [4] proposed the basis for the GAN and then many of the researchers worked on it, suggested some changes in the network, applied to various segments and then deep learning started to be applied to the underwater environment. Neural network algorithms, which were based on GAN [5–9], were then applied to the underwater image enhancement.

Various studies have used and modified the GAN model for bringing the performance improvisation underwater image enhancement, algorithms like underwater generative adversarial network (UWGAN) [10], WaterGAN [11], UGAN [12] and fast underwater image enhancement generative adversarial network (FUnIE-GAN) [13].

In 2017, Li et al. [11] presented WaterGAN, a generative adversarial network for underwater images. The function of the WaterGAN is that it uses real-world images and depth maps to synthesize them to the images underwater, and then creates the dataset of synthesized underwater images and creates the colour correction mechanism.

In 2018, Fabbri et al. [12] presented an adversarial network as the colour reconstruction technique named underwater GAN (UGAN). UGAN was trained on the dataset they have prepared using the CycleGAN. The CycleGAN was utilized to prepare the paired dataset as the style transformation from airy images to underwater images. It works as the generation of distorted underwater images from clear underwater (underwater images with no distortion) images. To train UGAN, the objective functions are similar to Wasserstein GAN (WGAN) formulation along with gradient difference loss between underwater image and generated image. In 2019, Guo et. al [10] proposed UWGAN, a new underwater GAN. The UWGAN is used to correct the image colour and retains the image details, the generator of this method is developed with the use of residual multi-scale dense block (RMSDB) in its encoder and decoder stages.

In later years, another method FUnIE-GAN [13] was proposed, which is fast in terms of time to generate enhanced coloured images. This method can be more suitable to be used with the remotely operated vehicle that are collecting data in real-time. It provides competitive performance for underwater image enhancement.

## 12.3 Proposed Method

The proposed architecture is shown in Figure 12.2 and is inspired by FUnie-GAN. It has two segments of architecture such as generator and discriminator. Fake images are

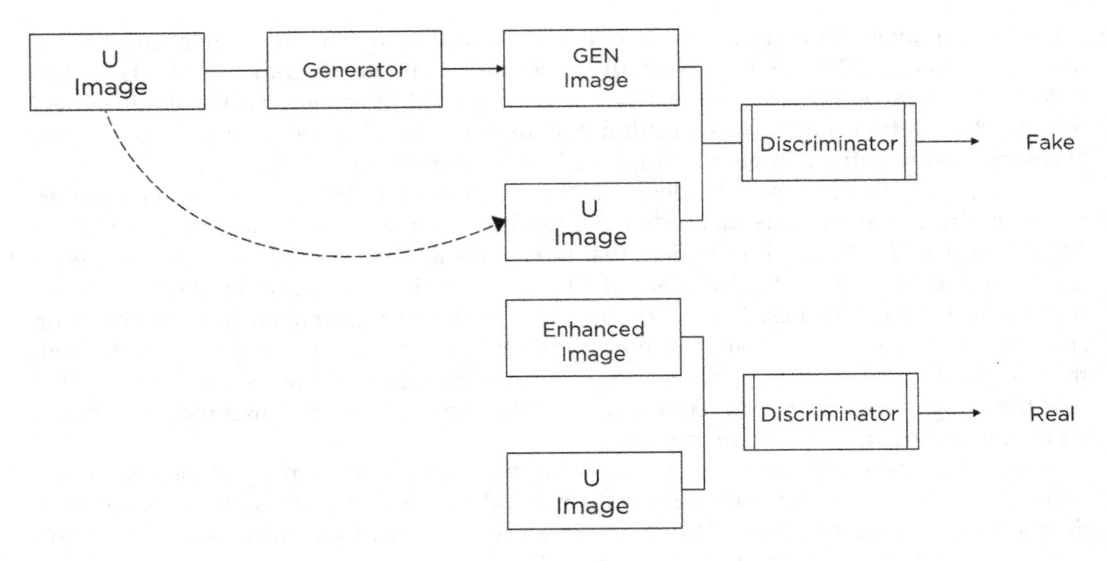


**FIGURE 12.2**
The proposed method for underwater image enhancement.

generated using the generator and for classification purposes such as whether the images are fake or real, we use a discriminator.

*Generator*: The generator in GAN adapts to generate fake underwater images by adding supervision from the discriminator architecture. Then discriminator tries to classify its generated output as real. Generators create data instances from the random data and whereas discriminators will classify the generated data and output with the loss [13]. We have utilized the U_Net architecture to develop the generator using the atrous-convolutions. In the encoder of u-net, the first two convolutional layers have been kept and then the Maxpooling operation is implemented, which reduces the feature maps received from the second convolutional layers both having 64 filters. The reduced features are convolved with two conjugative layers with the number of filters as 128, followed by a Maxpooling layer. Then the atrous convolution is applied up to the latent space in between multiple Maxpooling and atrous convolutional layers are there. The detailed generator architecture is shown in Figure 12.3.

The discriminator network in a GAN is a type of classifier as shown in Figure 12.4. It tries to distinguish enhanced images (labelled images) from the generated images. It could be of any network architecture depending on the available data type it is trying to classify. The training data of the discriminator belongs to two sources: real data instances and enhanced images, available in the dataset. The discriminator takes these images as positive class examples during training. Fake data instances are generated by the generator of GAN architecture. During training, these images (generated images) are treated as negative instances. The discriminators training procedure is well defined in the proposed algorithm. When an underwater image and the generated images are given to the discriminator, it is expected to classify the fake images. And, if the enhanced image and the underwater image are applied to the discriminator, then it is expected to classify the real image.

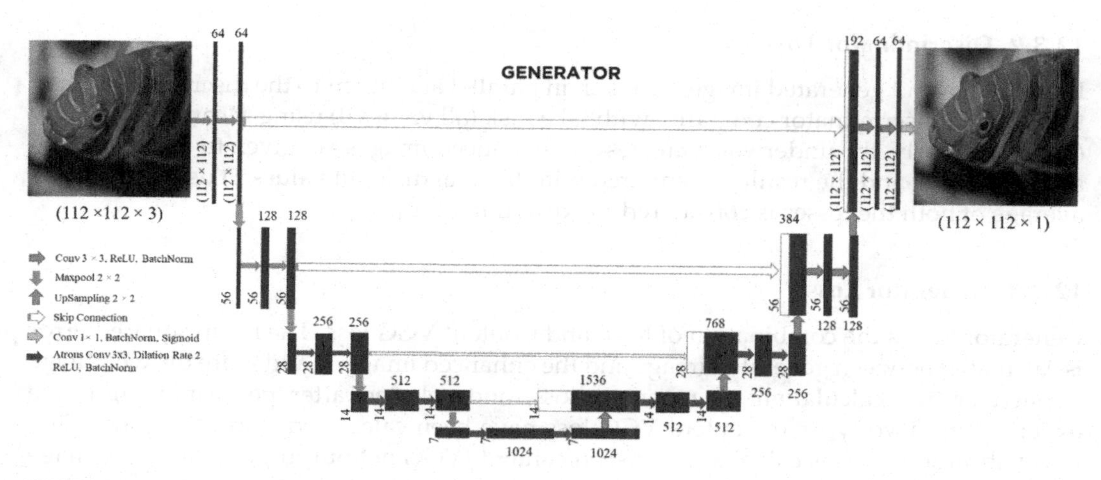


**FIGURE 12.3**
Generator of the proposed architecture.

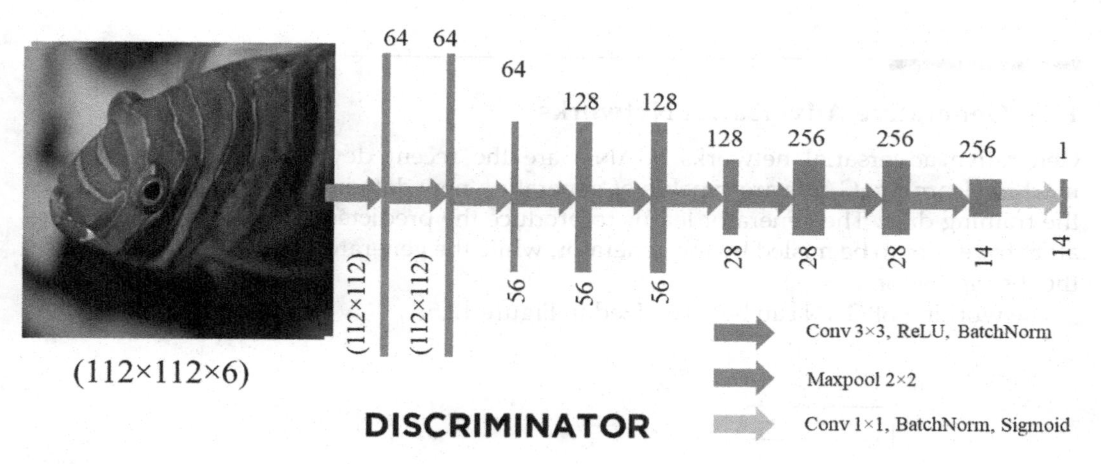


**FIGURE 12.4**
Discriminator of the proposed architecture.

### 12.3.1 Loss Function

$$L_D = Avg\left[MSE\left\{D\left(U_{img}, Gen_{img}\right), Fake_{disk}\right\}, MSE\left\{D\left(U_{img}, E_{img}\right), Real_{disk}\right\}\right] \quad (12.1)$$

$$L_G = \lambda_1 MSE\left\{G\left(U_{img}\right), E_{img}\right\} + \lambda_2 G_{loss}$$

$$G_{loss} = \alpha_1 \left\| Vgg\left\{G\left(U_{img}\right)\right\}, Vgg\left(E_{img}\right)\right\|_1 + \alpha_2 \left\| Vgg\left\{G\left(U_{img}\right)\right\}, Vgg\left(E_{img}\right)\right\|_2 \quad (12.2)$$

### 12.3.2 Discriminator Loss

Input image and generated image are taken in parallel and given to the discriminator than the result of discriminator compared with fake disk (all values 0) using Mean Square Error (MSE). Similarly, the underwater images and enhanced images are given to the discriminator in parallel and the result is compared with the real disk (all values 1) using MSE. The average of both the losses is considered the discriminator loss.

### 12.3.3 Generator Loss

Generator loss is the combination of MSE and Content VGG loss. The mean squared error is calculated between generated image and the enhanced image. $\lambda_1$and $\lambda_2$are the weightage applied on the calculation of generator loss, and selected after performing multiple experiments. Two types of content VGG loss have been calculated $L_1$and $L_2$. $L_1$ loss (first mean absolute error) is calculated between content (VGG net output) of generated image and enhanced image. Similarly, L2 loss is the Euclidian distance between the generated content and enhanced image content.

## 12.4 Generative Adversarial Networks

Generative adversarial networks (GANs) are the recent development in the field of machine learning. GANs are capable of generating new data instances that are similar to the training data. The generator learns to produce the predicted outcome. The discriminator strives not to be misled by the generator, while the generator tries not to be tricked by the discriminator.

The working of GAN can be visualized in Figure 12.5.

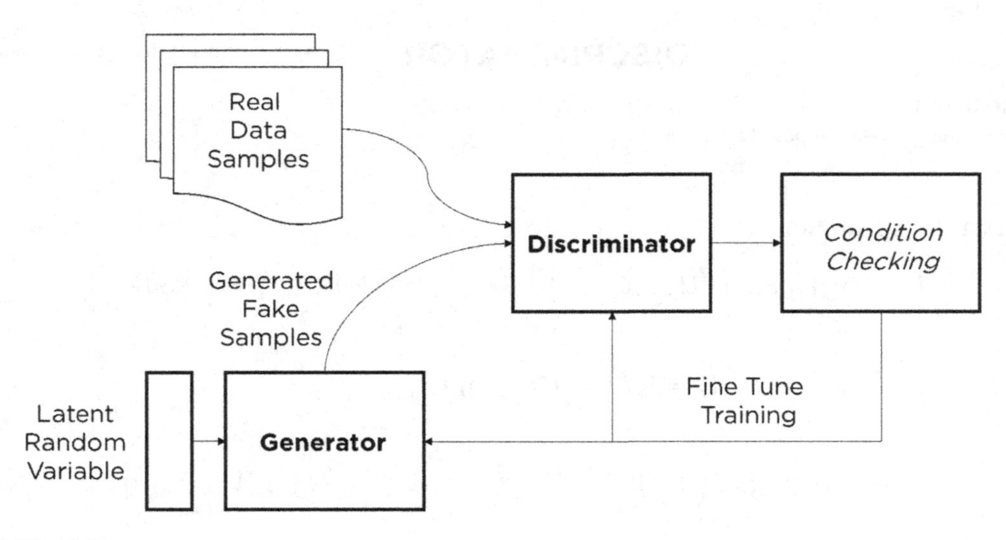


**FIGURE 12.5**
Function of generator and discriminator in GAN.

The generator records the data and trains itself in such a way that it will increase the probability of the discriminator is making mistake whereas the discriminator model estimates the probability that is data received from training, not from the generator.

The GANs are formulated as shown below [15]:

$$\min_{G}\max_{D} V(D,G)V(D,G) = E_{x\sim p_{data}(x)}\left[\log D(x)\right] + E_{z\sim p_z(z)}[\log\log\,(1 - D(G(z))]] \quad (12.3)$$

where

$$D = \text{Discriminator}$$

$$D(x) = \text{Discriminator network}$$

$$G = \text{Generator}$$

$$Pdata(x) = \text{Distribution of real data}$$

$$x = \text{sample from } Pdata(x)$$

$$P(z) = \text{distribution of generator}$$

$$z = \text{sample from } P(z)$$

$$G(z) = \text{Generator network}$$

## 12.5 Atrous Convolution

Dilated convolution is also known as atrous convolution or convolution with holes. Inflating the kernel means it will then skip some of the pixels based on the rate decided. We can see the difference between discrete convolution and atrous convolution. The discrete convolution can be written as:

$$(F * k)(p) = \sum_{s+t=p} F(s)k(t) \quad (12.4)$$

As kernel skips a few points based on the dilation rate, dilated convolution is written as

$$(F *_l k)(p) = \sum_{s+lt=p} F(s)k(t) \quad (12.5)$$

We can easily identify the difference between the discrete and atrous convolution from the summation it is evident. From Figure 12.5, we can easily determine the difference between the convolutions.

Figures 12.6 and 12.7 represent the images for dilated or atrous convolution wherein the pixels represented in red dot images we acquire after convolution. It can be observed the

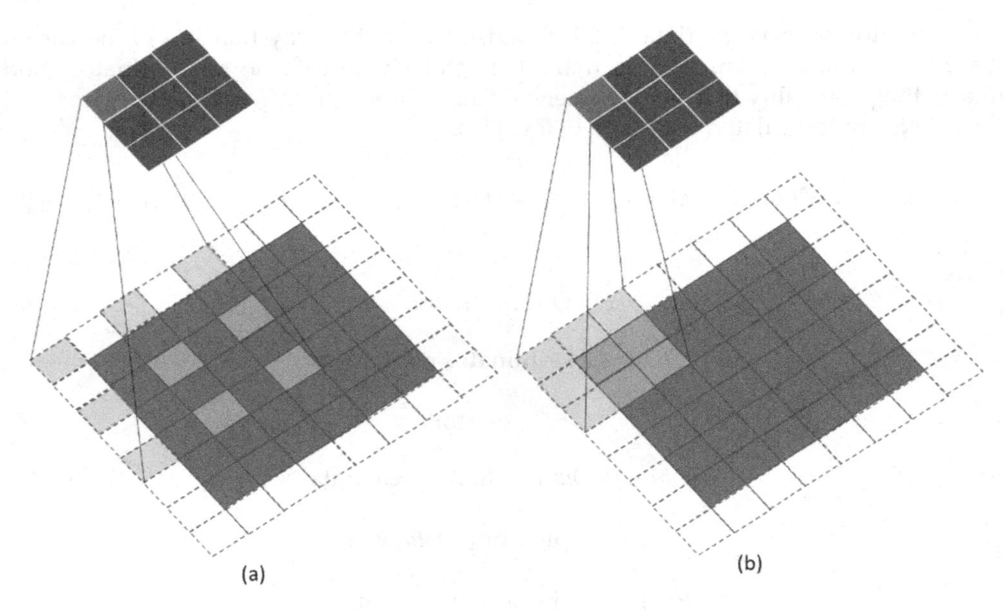


**FIGURE 12.6**
(a) Dilated convolution; (b) Discrete convolution.

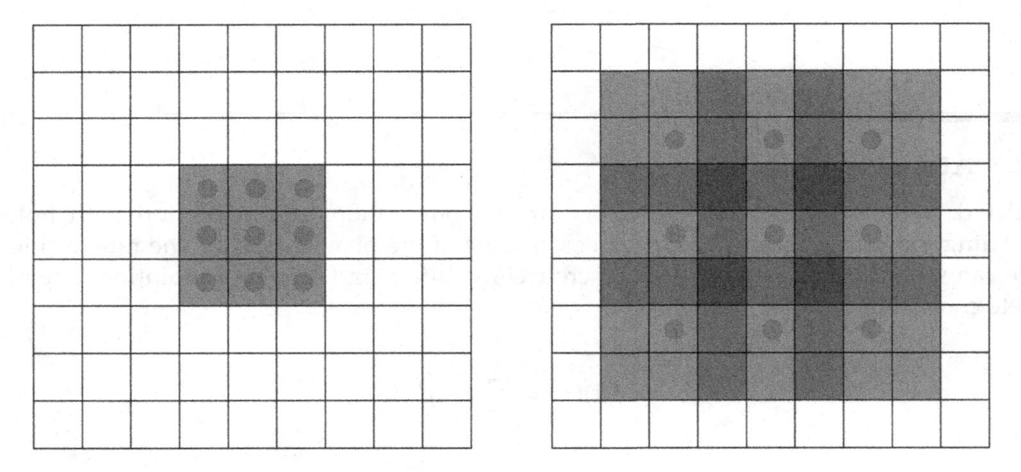

**FIGURE 12.7**
Dilated convolutions with different dilation rates.

difference between the images in the figure is that even though there are equal dimensions the receptive field is different, which is actually a dilated convolution dilation rate. So we can justify that increase in the receptive field represents more coverage of the image without adding additional costs.

## 12.6 Experimental Results

Objective evaluation can be done based on the statistics of the results, and for that, we have used performance evaluation parameters [16] such as peak signal-to-noise ratio (*PSNR*) [17], structural similarity index measure (*SSIM*) [18,20], the underwater image quality metric (*UIQM*) [18,19,20].

*Peak signal-to-noise ratio (PSNR)*: PSNR is the ratio of the image data power of a frame to the power of noise in the same frame. It measures the quality of the generated image to the reference image. To calculate the PSNR of an image, one must be able to compare that image to the enhanced image with the maximum power [21,23,25].

PSNR is defined as follows:

$$PSNR = 10log_{10}\left(\frac{(L-1)^2}{MSE}\right) = 20log_{10}\left(\frac{L-1}{RMSE}\right) \tag{12.6}$$

*Structural similarity index* (*SSIM*): The structural similarity index (SSIM) is an intuitive metric that assesses the image quality degradation that is caused by processing like data compression or loss caused in the transmission of data. This method requires two images for the analysis. SSIM index provides an approximation of human perceptual image quality. SSIM attempts to measure the change in three components, that is, luminance, contrast and structure in an image [21,25].

$$SSIM(x,y) = \frac{\left(2\mu_x\mu_y + c_1\right)\left(2\sigma_{xy} + c_2\right)}{\left(\mu_x^2 + \mu_y^2 + c_1\right)\left(\sigma_x^2 + \sigma_y^2 + c_2\right)} \tag{12.7}$$

### 12.6.1 Underwater Image Quality Measure (UIQM)

In underwater environment, it has been observed that due to different components like reflection, refraction and scattering, images can be a linear combination of all these components. Even these components are impacting colour, contrast etc. degradation. Therefore, its use of linear superposition model can be used for quality assessment. The overall UIQM is then given by

$$UIQM = c_1 \times UICM + c_2 \times UISM + c_3 \times UIConM \tag{12.8}$$

where the colour, contrast and sharpness measures are linearly added together. It is worth noting that the UIQM in [15,25] has three parameters. $c_1$, $c_2$, and $c_3$ above mentioned are the evaluation parameter, which is usually used while we analyse the image objectively, but apart from that mean square error (MSE), patch-based contrast quality index (PCQI), entropy and underwater colour image quality evaluation (UCIQE) are some of the other evaluation methods used in underwater image processing [20]. The selections of quality evaluation metrics or parameters are totally application dependent as shown in Table 12.1.

We have compared the results with few standard techniques and two recently published techniques. Histogram equalization (HE) is a method of contrast adjustment based on the histogram of the image. HE is used to spreading the image pixel values in the entire range

**TABLE 12.1**
Performance Evaluation Parameter on Input, Output Using U-UIQM and E-UIQM

| | | |
|---|---|---|
| **U-UIQM** | : | 2.56±0.55 |
| **E-UIQM** | : | 2.89±0.53 |
| **U-E-PSNR** | : | 19.96±2.88 |
| **U-E-SSIM** | : | 0.78±0.07 |

U-UIQM: Underwater (input image) UIQM
E-UIQM: Enhanced (output image) UIQM
U-E-PSNR: PSNR between Underwater & Enhanced image
U-E-SSIM: SSIM between Underwater & Enhanced image

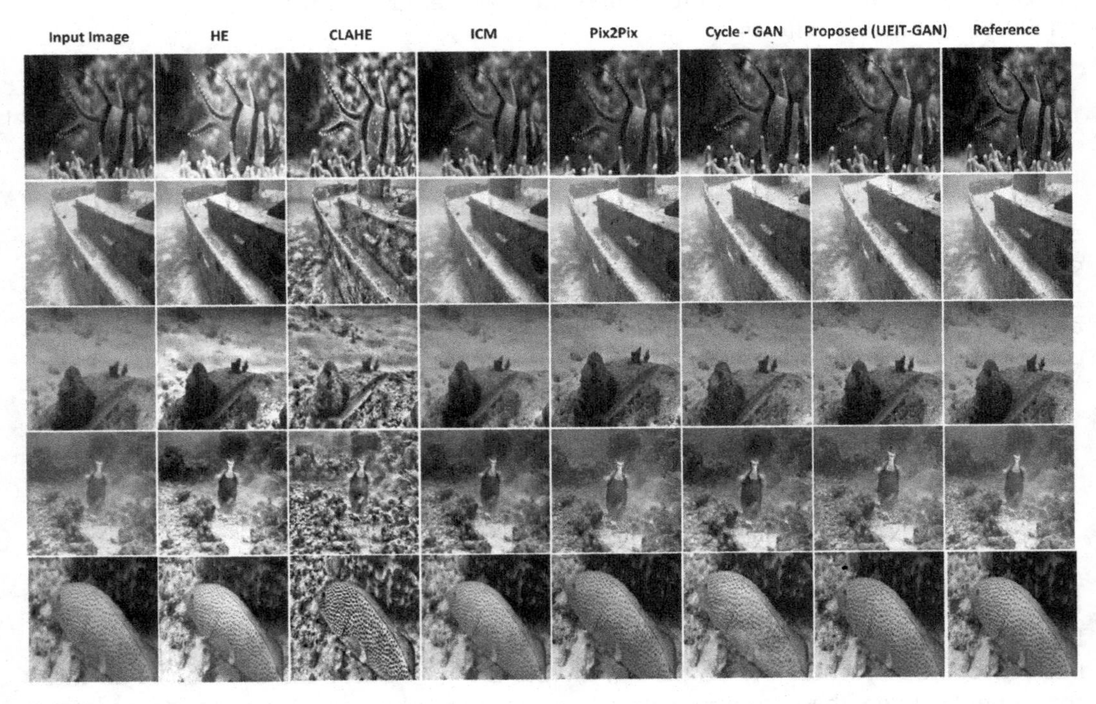


**FIGURE 12.8**
Subjective evaluation comparison of the proposed approach with various existing methods.

of 0 to 255 instead of having a fixed range of the underwater image. Underwater images are having low contrast, and image histogram equalization is able to fix the contrast problem, but it introduces the luminesce in the image. Figure 12.8 shows subjective evaluation comparison of the proposed approach with various existing methods.

Contrast limited adaptive histogram equalization (CLAHE) utilizes the contrast limiting procedure for each neighbourhood for which a transformation function is acquired. CLAHE was developed to avoid the over-amplification of the noise that is generated by the use of the histogram technique [22,25].

Underwater image enhancement using an integrated colour model, which then improves the perception of underwater images, The Pix2Pix GAN is used for image-to-image translation. which is designed based on conditional GAN, where a conditional target image is

**TABLE 12.2**
Performance Evaluation Parameter on an Input, Output Using P-UIQM

| | HE | CLAHE | ICM | Pix2Pix | CycleGAN | Proposed |
|---|---|---|---|---|---|---|
| P-UIQM | 2.87±0.48 | 3.24±0.27 | 2.66±0.54 | 3.38±0.26 | 3.51±0.18 | 3.35±0.24 |
| U-P-PSNR | 12.89±2.31 | 13.72±1.28 | 27.42±7.12 | 20.51±2.14 | 18.07±2.94 | 21.05±2.03 |
| E-P-PSNR | 13.66±2.68 | 13.84±1.61 | 20.67±2.93 | 21.51±2.87 | 19.66±2.34 | 24.96±3.48 |
| U-P-SSIM | 0.67±0.12 | 0.51±0.06 | 0.96±0.03 | 0.72±0.07 | 0.73±0.07 | 0.87±0.04 |
| E-P-SSIM | 0.52±0.12 | 0.51±0.07 | 0.80±0.6 | 0.65±0.08 | 0.70±0.07 | 0.81±0.07 |

P-UIQM: Proposed algorithm UIQM
U-P-PSNR: PSNR between Underwater & Proposed (resulted) image
E-P-PSNR: PSNR between Enhanced & Proposed (resulted) image
U-P-SSIM: SSIM between Underwater & Proposed (resulted) image
E-P-SSIM: SSIM between Enhanced & Proposed (resulted) image

generated [23]. CycleGAN is a model that addressed the problem and solves the image-to-image translation [24].

It is evident from the results that the proposed technique outperforms in both the evaluation mechanism, subjective and objective as depicted in Table 12.2.

## 12.7 Conclusions

Due to low light intensity and other parameters like diffusion and absorption of light and loss of colour spectrum wavelength, underwater images get degraded in their colour. In this chapter, we have presented the GAN-based method for the enhancement in terms of the colour of the underwater images. The utilization of deep learning technology has significantly increased due to its capability to solve complex problems and can determine the hidden pattern in data. Underwater images (corrupted) to enhance image translation technique is based on GAN, which includes the generator network and discriminator network. A generator network is developed with the inclusion of atrous convolution on UNET architecture. The proposed model formulates a generator loss function by evaluating image quality based on its global colour, content, local texture and style information, and discriminator loss is calculated using the mean MSE. We have represented the results in a qualitative and quantitative manner, and based on the results represented in this chapter, and from the results, it is evident that the presented model outperforms all other techniques. In future, real-time application of this method in ROV can be done with some of the applications, which can allow to improvise the probability of detection and identification with enhancement in real time.

## References

1. Wang, Y.; Zhang, J.; Cao, Y.; Wang Z. A deep CNN method for underwater image enhancement. In 2017 IEEE International Conference on Image Processing (ICIP) 2017 Sep 17 (pp. 1382–1386). IEEE.

2. Li, Xiu, et al. Fast accurate fish detection and recognition of underwater images with fast R-CNN. OCEANS 2015-MTS/IEEE Washington. IEEE, 2015.
3. Han, F.; Yao, J.; Zhu, H.; Wang C. Underwater image processing and object detection based on deep CNN method. *Journal of Sensors.* 2020 May 22, 2020, 1–20.
4. Goodfellow, I.; Pouget-Abadie, J.; Mirza, M.; Bing, X.; Bengio, Y. Generative Adversarial Nets. In Proceedings of the 27th International Conference on Neural Information Processing Systems, Montreal, QC, Canada, 8 December 2014; pp. 2672–2680.
5. Chen, B.; Xia, M.; Huang, J. MFANet: A Multi-level feature aggregation network for semantic segmentation of land cover. *Remote Sens.* 2021, 13, 731.
6. Xia, M.;Wang, T.; Zhang, Y.; Liu, J.; Xu, Y. Cloud/shadow segmentation based on global attention feature fusion residual network for remote sensing imagery. *Int. J. Remote Sens.* 2021, 42, 2022–2045.
7. Xia, M.; Cui, Y.; Zhang, Y.; Xu, Y.; Xu, Y. DAU-Net: A novel water areas segmentation structure for remote sensing image. *Int. J. Remote Sens.* 2021, 42, 2594–2621.
8. Xia, M.; Liu, W. Non-intrusive load disaggregation based on composite deep long short-term memory network. *Expert Syst. Appl.* 2020, 160, 113669.
9. Xia, M.; Zhang, X.; Liu, W.; Weng, L.; Xu, Y. Multi-stage feature constraints learning for age estimation. *IEEE Trans. Inf. Forensics Secur.* 2020, 15, 2417–2428.
10. Arjovsky, M.; Bottou L. Towards principled methods for training generative adversarial networks. arXiv preprint arXiv:1701.04862. 2017, Jan 17.
11. Li, J.; Skinner, K.; Eustice, R.; Johnson-Roberson, M. WaterGAN: Unsupervised generative network to enable real-time colour correction of monocular underwater images. *IEEE Robot. Autom. Lett.* 2017, 3, 387–394.
12. Fabbri, C.; Islam, M.; Sattar, J. Enhancing underwater imagery using generative adversarial networks. In Proceedings of the 2018 IEEE International Conference on Robotics and Automation, Brisbane, QLD, Australia, 21–25 May 2018; pp. 7159–7165.
13. Islam, M.; Xia, Y.; Sattar, J. Fast underwater image enhancement for improved visual perception. *IEEE Robot. Autom. Lett.* 2020, 5, 3227–3234.
14. Hu, Kai; Yanwen Zhang; Chenghang Weng; Pengsheng Wang; Zhiliang Deng; Yunping Liu. An underwater image enhancement algorithm based on generative adversarial network and natural image quality evaluation index. *J. Marine Sci. Eng.* 2021, 9(7), 691.
15. Zhao, X.; Jin, T.; Qu, S.. Deriving inherent optical properties from the background colour and underwater image enhancement. *Ocean Engineering*, 2015, 94, 163–172.
16. Jamadandi, A.; Mudenagudi, U.. Exemplar-based Underwater Image Enhancement Augmented by Wavelet Corrected Transforms. In Proceedings of the IEEE Conference on Computer Vision and Pattern Recognition Workshops, 2019, pp. 11–17.
17. Tang, C.; von Lukas, U.F.; Vahl, M.; Wang, S.; Wang, Y.; Tan, M. Efficient underwater image and video enhancement based on Retinex. *Signal, Image and Video Processing*, 2019, 13(5), 1011–1018.
18. Boom, B.J.; He, J.; Palazzo, S.; Huang, P.X.; Chou, Hsiu-Mei; Lin, Fang-Pang; Spampinato, C.; Fisher, R.B. A research tool for long-term and continuous analysis of fish assemblage in coral-reefs using underwater camera footage. *Ecol. Informatics*, 2014, DOI: dx.doi.org/10.1016/j.ecoinf.2013.10.006.
19. Yussof, W.N.; Hitam, M.S.; Awalludin, E.A.; Bachok Z. Performing contrast limited adaptive histogram equalization technique on combined colour models for underwater image enhancement. *International Journal of Interactive Digital Media.* 2013, 1(1), 1–6.
20. Emberton, S.; Chittka, L.; Cavallaro, A. Underwater image and video dehazing with pure haze region segmentation. *Computer Vision Image Understanding* 2018, 168, 145–156.
21. Islam, Md Jahidul; Youya Xia; Sattar, J. Fast underwater image enhancement for improved visual perception. *IEEE Robotics Automation Lett.* 2020, 5(2), 3227–3234.
22. Reza, Ali M. Realization of the contrast limited adaptive histogram equalization (CLAHE) for real-time image enhancement. J. VLSI Signal Processing Systems Signal, Image Video Technol. 2004, 38(1), 35–44.

23. Qu, Yanyun; Yizi Chen; Jingying Huang; Yuan Xie. Enhanced pix2pix dehazing network. In Proceedings of the IEEE/CVF Conference on Computer Vision and Pattern Recognition, 2019, pp. 8160–8168.
24. Fabbri, C.; Islam, Md J.; Sattar, J. Enhancing underwater imagery using generative adversarial networks. In 2018 IEEE International Conference on Robotics and Automation (ICRA), 2018, pp. 7159–7165. IEEE,.
25. Raveendran, S.; Patil, M. D.; Birajdar, G. K. Underwater image enhancement: a comprehensive review, of recent trends, challenges and applications. *Artificial Intelligence Rev.*, 2021, 54(7), 5413–5467.

# 13

# *Towards GAN Challenges and Its Optimal Solutions*

**Harmeet Kaur Khanuja and Aarti Amod Agarkar**

**CONTENTS**



## 13.1 Introduction: Background and Driving Forces

Generative models use an unsupervised learning approach (Goodfellow 2020). There are samples in the data that are input variables but it lacks the output variable in a generative model. The generative model is trained using only the input variables and an unknown output is generated based on training data on identifying the patterns from the input variables. We are more associated with generating the predictive models as done in supervised learning with the input variables; this type of classification model can be termed a discriminative model. The classification algorithm classifies or discriminates the examples into different classes as per their labels or categories, whereas the unsupervised algorithm creates or generates new examples as per the input data distribution. Two famous generative algorithms are autoencoders (Makhzani 2015) and generative adversarial networks (Goodfellow 2014). Autoencoders are one of the types of artificial neural networks that draw efficient data coding to bring out the essence and meaning of the data provided in an unsupervised manner. An autoencoder intends to discover a depiction (encoding) for the input data provided, usually for dimensionality reduction. Generative adversarial networks (GANs) are another set of generative algorithms which can generate images that can make people speculate if it is a real image or an image generated by GANs. GANs comprise two neural networks that work together; they are called as generator and discriminator, which follow the unsupervised learning approach. GANs are trained using neural networks by supplying deceptive input which is called adversarial setting. Figure 13.1 shows the system architecture of GANs. The aim of a generator is to generate

DOI: 10.1201/9781003203964-13

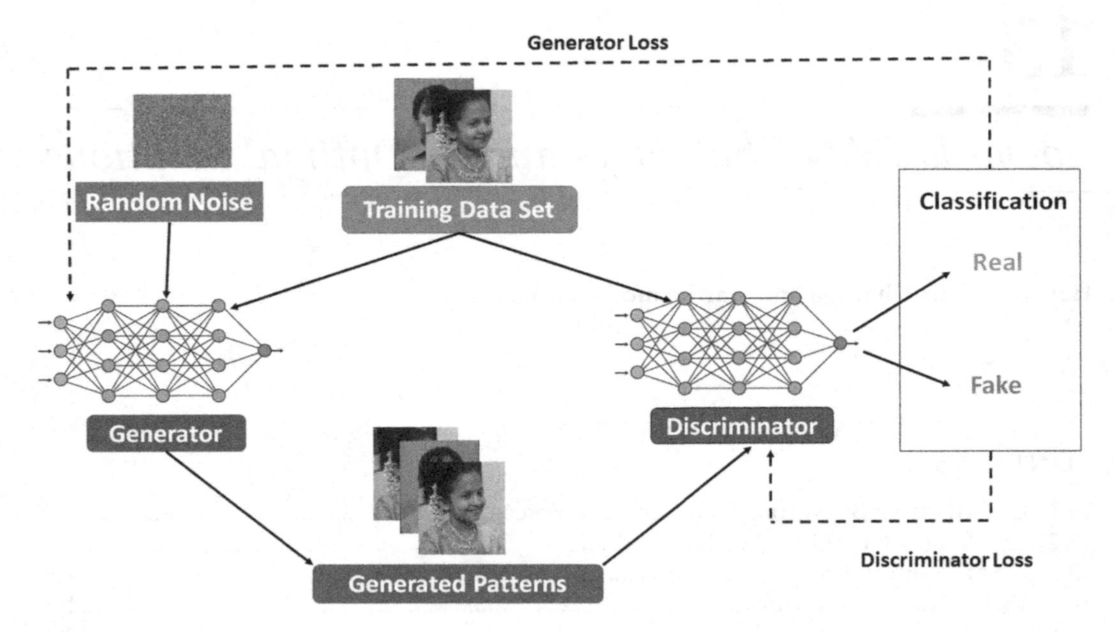


**FIGURE 13.1**
System architecture of generative adversarial networks.

data for example a fake picture that looks like a real picture while the aim of the discriminator is to look at the generated data from generator and discriminate it as real or fake.

GANs meet various failure situations, which raises challenges to build a trustworthy model. The common problems encountered boost active research areas. This chapter describes problems and various solutions to overcome challenges of GAN.

## 13.2 Challenges with GAN

GANs have received excellent success (Alqahtani 2021) but at the same time there are still challenges to make the GAN model stable. The problems are majorly associated during training GAN model due to Nashequilibrium, vanishing gradient, mode collapse and non-convergence issues. Thus, stable training is a significant issue in the applications for the successful GAN. In this chapter we discussed distinct research challenges during training GANs and their solutions as suggested by different researchers, which provide solutions for sustainable GAN training. This chapter discusses the various GAN training problems like Nashequilibrium, vanishing gradient, mode collapse and non-convergence issues and their corresponding solutions. This gives new directions for researchers.

## 13.3 GAN Training Problems

Goodfellow et al. (2014) presented the concept of GAN. In this work, generator and discriminator are presented as players of a non-cooperative game model. In this work, both generator and discriminator are two deep neural networks. Here, generator is trained to generate different patterns. Discriminator has both the inputs; viz. real patterns and generated patterns. The discriminator attempts to distinguish between real and generated pictures whereas the generator attempts to fool the discriminator with generating patterns near to real patterns. The GAN training problems that occur are discussed in the further sections.

### 13.3.1 NashEquilibrium

Nashequilibrium is the concept from game theory. Two-player game from the game theory is a zero-sum problem where two players play with each other and try to gain the benefit or win the match by defeating the opponent. When one player wins the game, the other loses as per the zero-sum principle. Both players take the decision by considering the possible decisions of the opponent. It is an example of an optimization problem where both players try to win. During the decision making of both the opponents, it may be possible that both players reach the situation where they gain maximum possible benefit. If any player makes the next decision, there is no chance to have better gain. This situation is called the Nashequilibrium state. Figure 13.2 shows the conceptual diagram of the Nashequilibrium state where both the generator model and discriminator model are at the optimal state.

In GAN, generator and discriminator are two players and compete with each other. If the generator generates accurate fake images, then it is difficult for the discriminator to recognize them. If the discriminator is trained well and can discriminate between real and fake images accurately, then it can be considered as the limitation of the generator. With the process of training generator and discriminator, one situation can be reached, which is an optimized situation for generator as well as discriminator. In this state, the generator and discriminator give the best possible outcome. This is the Nashequilibrium state of GAN.

The survey on GAN literature shows that the stability of GAN (Arjovsky and Bottou 2017) is affected due to Nash equilibrium (Kurach 2019). It describes a situation where both generator G and discriminator D give their best performance. This is one of the objectives of GANs to reach a Nash equilibrium stage, which improves their individual performance. But ideally it is very difficult to reach at this stage because both the networks lack in communicating between them, which cannot update their cost functions independently. The

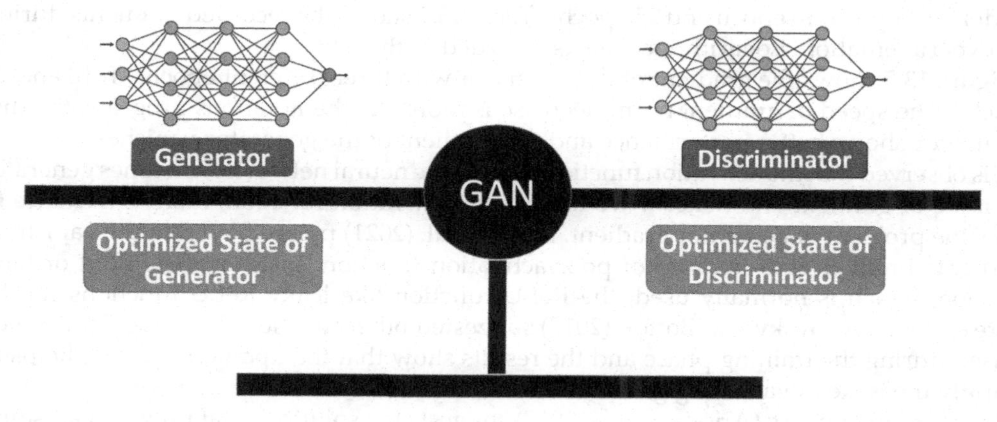


**FIGURE 13.2**
Nash equilibrium state.

GAN variants such as the WGAN (Arjovsky and Bottou 2017) ) and two time-scale update rule (TTUR) may be a solution to reach the Nash equilibrium (Heusel et al. 2017).

### 13.3.2 Vanishing Gradient

As mentioned previously, generator and discriminator models try to win against each other during GAN training. The vanishing gradient problem is the effect of the optimal discriminator model, which can accurately discriminate among real and fake examples at the training process. One of the outputs of the discriminator model is given as the feedback to the generator model, which is referred as generator loss. When the discriminator model perfectly discriminates between real and fake examples during the training phase, the generator receives no gradient and thus cannot improve itself during the generator's training phase. The vanishing gradient problem was detected by Denton et al. (2015), Arjovsky and Bottou (2017), Kurach et al. (2019), and Che et al. (2016). As the discriminator shows the best possible performance, the discriminator does not give feedback to the generator to improve.

It is experimented in the research paper (Arjovsky and Bottou 2017), if the supports of both the distributions are continuous then the model always obtains a perfect discriminator, which leads to an unreliable training of the generator by leading to vanishing gradient problem. Two experiments were carried out to show why the vanishing gradient problem occurs.

The first experiment is done for recording the discriminator's accuracy while training. As part of training the GAN model, the generator was trained for 1 epoch, 10 epochs and 25 epochs of the deep convolutional generative adversarial network, which is also called as DCGAN model. After training the generator model, the discriminator model was trained. As shown in Table 13.1, the cross-entropy was recorded for a number of iterations.
Figure 13.3 shows the discriminator's error during these iterations.

Figure 13.4 shows the discriminator's accuracy during these training iterations. For 1 epoch, 10 epochs and 25 epochs, the discriminator model shows good performance and can easily categorize the real and fake images. Within initial 50 to 100 iterations, the discriminator model becomes mature and can easily discriminate between original and generated images. This prevents the process of further improvement of the discriminator and thus the gradients.

The second experimentation is done for recording the gradient of the generator. As per the cost function defined in Arjovsky and Bottou (2017), we have trained our DCGAN model for 1 epoch, 10 epoch and 25 epochs. Table 13.2 shows the recorded gradients during the experimentation. Logarithmic scale is recorded in the table.

Figure 13.5 shows the gradient of the generator with three different epochs: 1, 10 and 25 epochs. The speed of gradient norm decrease is more. As the iterations progress, the discriminator shows better performance and the gradient of the generator vanishes.

It is observed that the activation function used in the neural network sometimes generates the issue of gradient and hence if we modify the activation function, there is a chance to solve the problem of generator gradient. Kladis et al. (2021) pointed out that the vanishing gradients problem was because of poor activation function. Instead of sigmoid or tanh function, which is normally used, the ReLU function like leaky ReLU functions can be more suitable. Arjovsky and Bottou (2017) suggested other functions like $-\log D(x)$, which is used during the training phase and the results show that the updated function helps to simplify the issue of vanishing gradient.

The Wasserstein loss (Arjovsky et al. 2017) suggests the solutions that prevent vanishing gradients even though the discriminator is trained at optimum level. Wasserstein distance

**TABLE 13.1**
Discriminator's Error

| | Cross-entropy | | |
|---|---|---|---|
| **Training Iterations** | **After 1 Epoch** | **After 10 Epoch** | **After 25 Epochs** |
| 10 | 1.914255925 | 52.11947111 | 0.008531001 |
| 100 | 0.056363766 | 1.294195841 | 0.000570164 |
| 200 | 0.013396767 | 0.106414302 | 0.00022856 |
| 300 | 0.005152286 | 0.038636698 | 0.000101158 |
| 400 | 0.002837919 | 0.017298164 | 7.99834E-05 |
| 500 | 0.002013724 | 0.013365955 | 4.96592E-05 |
| 600 | 0.001158777 | 0.007979947 | 2.85102E-05 |
| 700 | 0.000843335 | 0.006151769 | 2.20293E-05 |
| 800 | 0.000672977 | 0.004528976 | 1.63682E-05 |
| 900 | 0.000477529 | 0.003597493 | 1.34276E-05 |
| 1000 | 0.000380189 | 0.002710192 | 1.01158E-05 |
| 1200 | 0.000239883 | 0.001548817 | 7.43019E-06 |
| 1300 | 0.000185353 | 0.001324342 | 5.72796E-06 |
| 1400 | 0.000151356 | 0.001035142 | 4.7863E-06 |
| 1500 | 0.000135207 | 0.000820352 | 4.33511E-06 |
| 1600 | 0.00011749 | 0.000746449 | 4.00867E-06 |
| 1700 | 0.000104472 | 0.000626614 | 3.11889E-06 |
| 1800 | 8.41395E-05 | 0.000530884 | 2.84446E-06 |
| 1900 | 7.46449E-05 | 0.000450817 | 3.0761E-06 |
| 2000 | 7.06318E-05 | 0.000400867 | 2.69774E-06 |
| 2100 | 6.13762E-05 | 0.00036141 | 2.07491E-06 |
| 2200 | 4.69894E-05 | 0.000325087 | 1.70608E-06 |
| 2300 | 3.97192E-05 | 0.00027227 | 1.58855E-06 |
| 2400 | 3.59749E-05 | 0.00021928 | 1.3366E-06 |
| 2500 | 3.10456E-05 | 0.000190985 | 1.18577E-06 |
| 2600 | 2.73527E-05 | 0.000174181 | 1.13501E-06 |
| 2700 | 2.54097E-05 | 0.000159221 | 1.09396E-06 |
| 2800 | 2.44906E-05 | 0.000146218 | 1.05439E-06 |
| 2900 | 2.25944E-05 | 0.000132739 | 1.01859E-06 |
| 3000 | 2.06063E-05 | 0.00011885 | 9.28966E-07 |
| 3100 | 1.7378E-05 | 0.000108643 | 8.27942E-07 |
| 3200 | 1.44877E-05 | 9.977E-05 | 7.49894E-07 |
| 3300 | 1.2942E-05 | 9.07821E-05 | 6.53131E-07 |
| 3400 | 1.19124E-05 | 8.62979E-05 | 5.74116E-07 |
| 3500 | 1.09901E-05 | 8.24138E-05 | 5.74116E-07 |
| 3600 | 1.01625E-05 | 7.8886E-05 | 5.29663E-07 |
| 3700 | 9.1622E-06 | 7.53356E-05 | 4.57088E-07 |
| 3800 | 8.3946E-06 | 6.9024E-05 | 4.19759E-07 |
| 3900 | 7.8886E-06 | 6.74528E-05 | 3.77572E-07 |
| 4000 | 7.8886E-06 | 6.9024E-05 | 3.0903E-07 |

or earth mover's (EM) distance is defined as a measure of distance between two probability distributions. Arjovsky et al. (2017) discussed that when the critic is trained at its optimal level the earth mover's distance remains continuous and also it is differentiable. As the critic is trained at an optimal level, a more reliable Wasserstein gradient is obtained. The discriminator then learns instantly to distinguish between fake and real samples with less reliable gradient information. It is observed that the critic does not reach a saturation point. Its gradients converge to a linear function. These discriminators learn to differentiate the Gaussians distribution. In contrast, it is seen that the discriminator of a minimax GAN

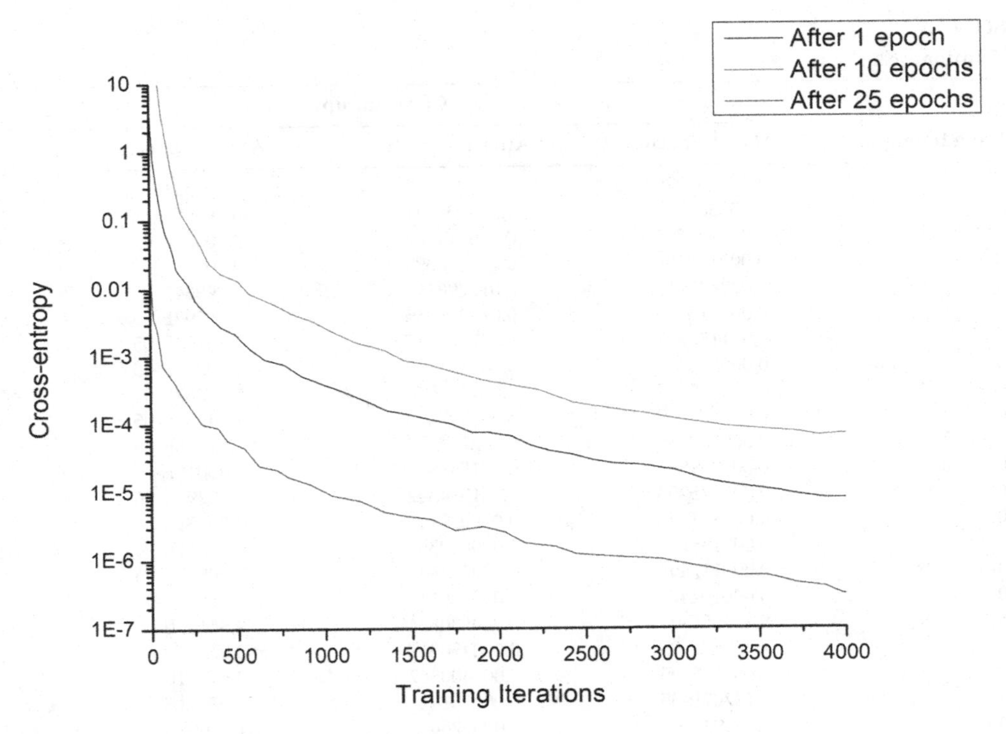


**FIGURE 13.3**
Discriminator's error.

saturates, resulting in vanishing gradients. But the Wasserstein GAN or WGAN critic still provides gradients that save the gradients from completely vanishing.

#### *13.3.2.1 Mode Collapse and Non-Convergence*

The aim of training phase of the generator and discriminator model is to reach the Nash equilibrium state. Vanishing gradient issue is one of the hurdles to achieve this objective. The problem of vanishing gradient can be reduced; one of the solutions is alternating gradient updates process (AGD). The process helps to solve the gradient problem, but at the same time introduces another issue, which is mode collapse (Bengio et al. 2017). In GAN, when the feedback of the discriminator is given as input to the generator, it may be possible that the generator or discriminator is stuck in local minima, which is referred to as the mode collapse problem.

In real time, the data distributions can be multi-modal as in the MNIST dataset, which have variation from digits "0" to "9", generated using GAN (Metz et al. 2016). We did an experiment where we generated MNIST images as shown in Figure 13.6. It was observed only the first row produces all the variation of ten modes whereas second row generates a single mode like digit "7". It is because of mode collapse problem where only a few modes of data are generated.

GAN must be trained on a large dataset to set the model to a higher quality model. Whenever the GAN is being trained for small dataset, the model faces two problems viz., mode collapse and non-convergence. During the mode collapse situation, when two

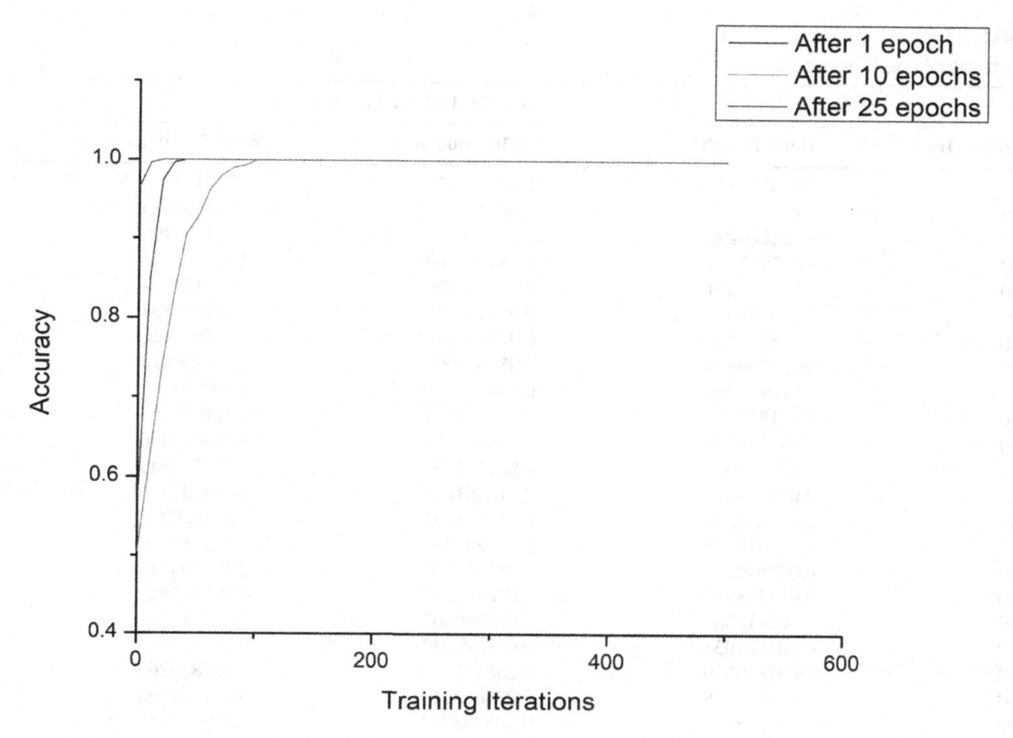


**FIGURE 13.4**
Discriminator's accuracy.

different inputs are given to the generator, it may be possible that the generator gives the same output. It reduces diversity in the output of the generator. It is pointed out in the research paper by Fedus et al. (2017), which gives a research challenge problem for a deeper investigation of GAN dynamics during training. How to increase the diversity of the generated samples remains a challenging problem.

One of the approaches to solve these two problems is gradient penalties presented by Gulrajani et al. (2017), Kodali et al. (2017) and which is experimented in Thanh-Tung et al. (2018). Gradient penalties help to improve the diversity of generated examples. When the real sample data distribution and generated sample data distribution are close to each other, it is difficult for the discriminator to precisely distinguish between real and fake samples. In Kodali et al. (2017) and Fedus et al. (2017), authors present that the gradient penalty helps to spread the fake examples over and near the real data samples. The process solves the problem of mode collapse and provides stability to the GAN model during the training phase.

In the experimentation using mini-batch, diverse fake examples are paired against different real examples where fake examples are attracted towards different real examples. One of the solutions proposed is to make batches of samples to increase the diversity of assessment. Mini-batch discrimination (MBD) in Salimans et al. (2016) gives the solution to this problem. Dissimilar mini-batches of real examples make the fake examples move in varied directions. This spreads fake examples above the real data distributions, which in effect try to reduce mode collapse. With this approach, the effect of mode collapse reduces, but it can make GAN reach anon-convergent state. Because the pair of fake and

**TABLE 13.2**
Gradient of the Generator

| Training Iterations | Gradient of the Generator | | |
|---|---|---|---|
| | After 1 Epoch | After 10 Epochs | After 25 Epochs |
| 10 | 6.776415076 | 21.92804935 | 6.776415076 |
| 100 | 0.497737085 | 3.60578643 | 0.497737085 |
| 200 | 0.134276496 | 0.680769359 | 0.134276496 |
| 300 | 0.063095734 | 0.336511569 | 0.063095734 |
| 400 | 0.033419504 | 0.148251809 | 0.033419504 |
| 500 | 0.024490632 | 0.092257143 | 0.024490632 |
| 600 | 0.015885467 | 0.072610596 | 0.015885467 |
| 700 | 0.012359474 | 0.057543994 | 0.012359474 |
| 800 | 0.010839269 | 0.045498806 | 0.010839269 |
| 900 | 0.011857687 | 0.034514374 | 0.011857687 |
| 1000 | 0.006039486 | 0.026977394 | 0.006039486 |
| 1100 | 0.006397348 | 0.02113489 | 0.006397348 |
| 1200 | 0.005333349 | 0.016481624 | 0.005333349 |
| 1300 | 0.003828247 | 0.014825181 | 0.003828247 |
| 1400 | 0.002910717 | 0.012647363 | 0.002910717 |
| 1500 | 0.002500345 | 0.011376273 | 0.002500345 |
| 1600 | 0.003318945 | 0.010447202 | 0.003318945 |
| 1700 | 0.002118361 | 0.008550667 | 0.002118361 |
| 1800 | 0.001524053 | 0.007379042 | 0.001524053 |
| 1900 | 0.001849269 | 0.006412096 | 0.001849269 |
| 2000 | 0.001210598 | 0.005623413 | 0.001210598 |
| 2100 | 0.001393157 | 0.005058247 | 0.001393157 |
| 2200 | 0.0012218 | 0.004634469 | 0.0012218 |
| 2300 | 0.001137627 | 0.004375221 | 0.001137627 |
| 2400 | 0.001002305 | 0.004355119 | 0.001002305 |
| 2500 | 0.000826038 | 0.004008667 | 0.000826038 |
| 2600 | 0.000726106 | 0.003491403 | 0.000726106 |
| 2700 | 0.000726106 | 0.003184198 | 0.000726106 |
| 2800 | 0.000726106 | 0.002864178 | 0.000726106 |
| 2900 | 0.000562341 | 0.00254683 | 0.000562341 |
| 3000 | 0.000504661 | 0.002328091 | 0.000504661 |
| 3100 | 0.000476431 | 0.002218196 | 0.000476431 |
| 3200 | 0.000406443 | 0.002123244 | 0.000406443 |
| 3300 | 0.000382825 | 0.002009093 | 0.000382825 |
| 3400 | 0.000389942 | 0.001819701 | 0.000389942 |
| 3500 | 0.000423643 | 0.001713957 | 0.000423643 |
| 3600 | 0.000472063 | 0.001667247 | 0.000472063 |
| 3700 | 0.000353997 | 0.001625549 | 0.000353997 |
| 3800 | 0.000265461 | 0.001285287 | 0.000265461 |
| 3900 | 0.000231206 | 0.001213389 | 0.000231206 |
| 4000 | 0.000244906 | 0.001291219 | 0.000244906 |

real examples are chosen at random, fake samples will be continually attracted towards different real data points during training and this will not lead the model to converge to any real samples. An optimal way shown in the paper by Salimans et al. (2016) is to perish the weight of the gradient penalty term.

Some regularization techniques are pointed to improve GAN convergence. It was experimented by Arjovsky and Bottou (2017) by adding noise to discriminator inputs and as suggested in Roth et al. (2017). They also penalize the discriminator weights. The

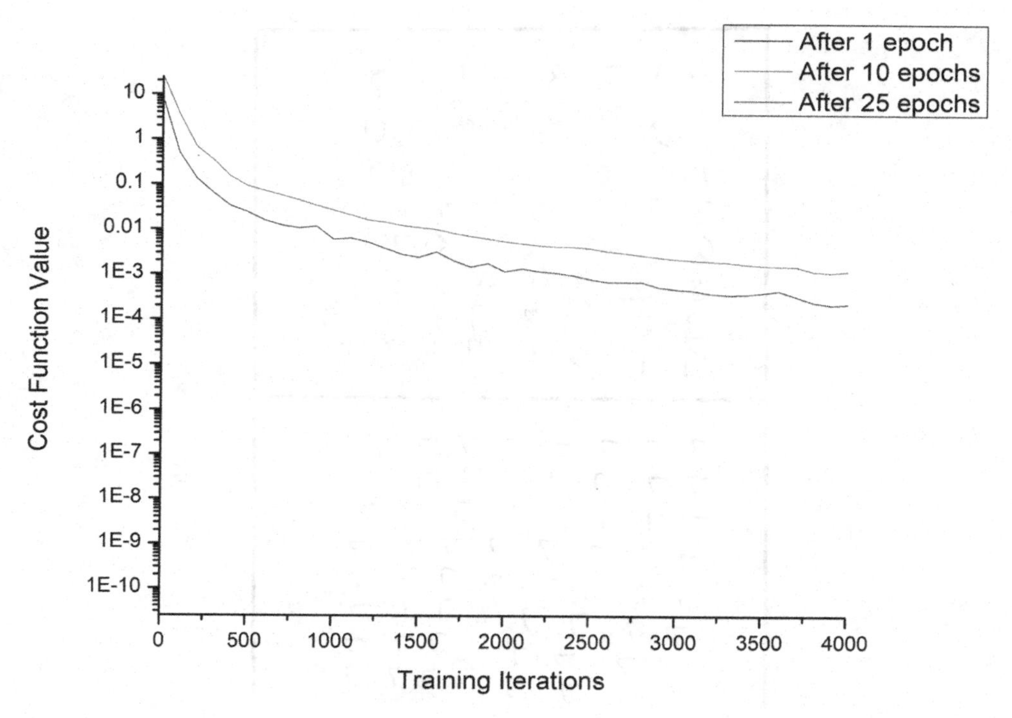


**FIGURE 13.5**
Gradient of the generator.

important parameter while training is to train the discriminator in such a way that it will not introduce the problem of vanishing gradient and at the same time it should give accurate feedback to the generator to modify the performance of the generator. If the discriminator is accurately trained, GAN shows the best possible performance. Unfortunately, there is no indicator of whether the discriminator is properly trained or not. WGAN (Arjovsky et al. 2017) and unrolled GAN (Metz et al. 2016) optimize the objective function and can overcome the problem as well.

## 13.4 Conclusion

The future of deep learning is generative adversarial networks. As GAN has extensive and gigantic applications in generating images, photographs of human faces, image-to-image translation and so on, this makes the GAN training challenges unavoidable. This chapter presents the basic model of GAN and the challenges with probable solutions experimented and proposed by various researchers for those challenges. As a solution, many GAN variants like WGAN, DCGAN, and CGANs are surveyed in Panet al. (2019) to overcome the challenges of GAN. This set forth the open research challenges for optimizing training methods of GAN moving towards a better GAN model.

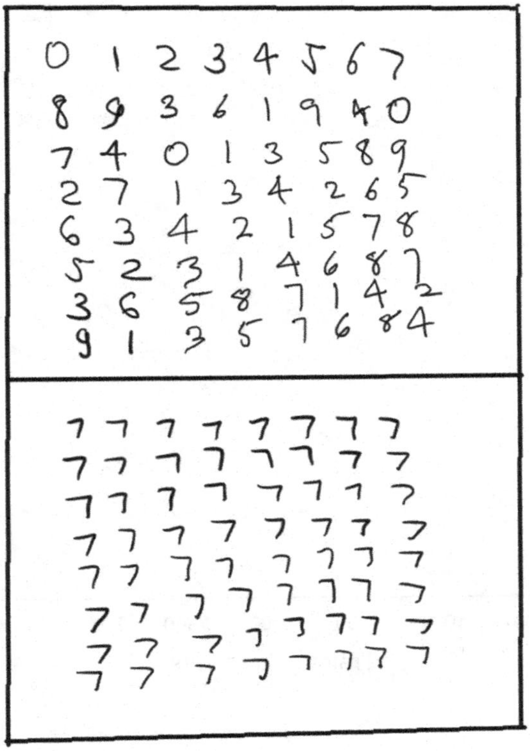

**FIGURE 13.6**
MNIST: GAN generated images.

## References

Alqahtani, H., Kavakli-Thorne, M., & Kumar, G. (2021). Applications of generative adversarial networks (GANs): An updated review. *Archives of Computational Methods in Engineering, 28*(2), 525–552.

Arjovsky, M., & Bottou, L. (2017). Towards principled methods for training generative adversarial networks. *arXiv preprint arXiv:1701.04862*.

Arjovsky, M., Chintala, S., & Bottou, L. (2017, July). Wasserstein generative adversarial networks. In *International conference on machine learning* (pp. 214–223). PMLR.

Bengio, Y., Goodfellow, I., & Courville, A. (2017). *Deep Learning*. Vol. 1. Cambridge, MA: MIT Press.

Che, T., Li, Y., Jacob, A. P., Bengio, Y., & Li, W. (2016). Mode regularized generative adversarial networks. *arXiv preprint arXiv:1612.02136*.

Denton, E. L., Chintala, S., & Fergus, R. (2015). Deep generative image models using aLaplacian pyramid of adversarial networks. *Advances in Neural Information Processing Systems28, NeurIPS Proceedings.* arXiv *preprint* arXiv:1506.05751

Fedus, W., Rosca, M., Lakshminarayanan, B., Dai, A. M., Mohamed, S., & Goodfellow, I. (2017). Many paths to equilibrium: GANs do not need to decrease a divergence at every step. *arXiv preprint arXiv:1710.08446*.

Goodfellow, I. (2017). GANs for Creativity and Design. www.iangoodfellow.com.

Goodfellow, I., Pouget-Abadie, J., Mirza, M., Xu, B., Warde-Farley, D., Ozair, S., ...& Bengio, Y. (2020). Generative adversarial networks. *Communications of the ACM, 63*(11), 139–144.

Goodfellow, I., Pouget-Abadie, J., Mirza, M., Xu, B., Warde-Farley, D., Ozair, S., ...& Bengio, Y. (2014). Generative adversarial nets. *Advances in Neural Information Processing Systems, 27, arXiv preprint arXiv:1406.2661*

Gulrajani, I., Ahmed, F., Arjovsky, M., Dumoulin, V., & Courville, A. (2017). Improved training of Wassersteingans. *arXiv preprint arXiv:1704.00028*.

Heusel, M., Ramsauer, H., Unterthiner, T., Nessler, B., & Hochreiter, S. (2017). GANs trained by a two time-scale update rule converge to a local Nash equilibrium. , *31st Conference on Neural Information Processing Systems (NIPS 2017), Long Beach, CA, USA.*

Kladis, E., Akasiadis, C., Michelioudakis, E., Alevizos, E., & Artikis, A. (2021). An empirical evaluation of early time-series classification algorithms. In *EDBT/ICDT Workshops*.

Kodali, N., Abernethy, J., Hays, J., & Kira, Z. (2017). On convergence and stability of gans. *arXiv preprint arXiv:1705.07215*.

Kurach, K., Lučić, M., Zhai, X., Michalski, M., & Gelly, S. (2019, May). A large-scale study on regularization and normalization in GANs. In *International Conference on Machine Learning* (pp. 3581–3590). PMLR.

Makhzani, A., Shlens, J., Jaitly, N., Goodfellow, I., & Frey, B. (2015). Adversarial autoencoders. *arXiv preprint arXiv:1511.05644*.

Metz, L., Poole, B., Pfau, D., & Sohl-Dickstein, J. (2016). Unrolled generative adversarial networks. *arXiv preprint arXiv:1611.02163*.

Pan, Z., Yu, W., Yi, X., Khan, A., Yuan, F. and Zheng, Y. (2019). Recent progress on generative adversarial networks (GANs): A survey. *IEEE Access, 7*:36322–36333.

Radford, A., Metz, L., & Chintala, S. (2015). Unsupervised representation learning with deep convolutional generative adversarial networks. *arXiv preprint arXiv:1511.06434*.

Roth, K., Lucchi, A., Nowozin, S., & Hofmann, T. (2017). Stabilizing training of generative adversarial networks through regularization. *arXiv preprint arXiv:1705.09367*.

Salimans, T., Goodfellow, I., Zaremba, W., Cheung, V., Radford, A., & Chen, X. (2016). Improved techniques for training gans. *Advances in Neural Information Processing Systems, 29*, 2234–2242.

Thanh-Tung, H., & Tran, T. (2018) On catastrophic forgetting and mode collapse in generative adversarial networks. www. groundai.com/project/on-catastrophic-forgetting-and-modecollapse-in-generative-adversarial-networks/1.

# Index

Printed in the United States
by Baker & Taylor Publisher Services